Archimedes

What Did He Do Besides Cry Eureka?

Cover Image: The Archimedes Palimpsest
Courtesy of Christie's Images,
Ltd. 1999. Used with permission.

Library of Congress Catalog Card Number 99-62795

ISBN 0-88385-718-9

Printed in the United States of America

Current Printing (last digit):
10 9 8 7 6 5 4 3

Archimedes

What Did He Do Besides Cry Eureka?

Sherman Stein
University of California, Davis

Published and Distributed by
THE MATHEMATICAL ASSOCIATION OF AMERICA

Classroom Resource Materials is intended to provide supplementary classroom material for students—laboratory exercises, projects, historical information, textbooks with unusual approaches for presenting mathematical ideas, career information, etc.

101 Careers in Mathematics, edited by Andrew Sterrett
Archimedes: What Did He Do Besides Cry Eureka? Sherman Stein
Calculus Mysteries and Thrillers, R. Grant Woods
Combinatorics: A Problem Oriented Approach, Daniel A. Marcus
A Course in Mathematical Modeling, Douglas Mooney and Randall Swift
Cryptological Mathematics, Robert Edward Lewand
Elementary Mathematical Models, Dan Kalman
Geometry from Africa: Mathematical and Educational Explorations, Paulus Gerdes
Interdisciplinary Lively Application Projects, edited by Chris Arney
Inverse Problems: Activities for Undergraduates, Charles W. Groetsch
Laboratory Experiences in Group Theory, Ellen Maycock Parker
Learn from the Masters, Frank Swetz, John Fauvel, Otto Bekken, Bengt Johansson, and Victor Katz
Mathematical Modeling in the Environment, Charles Hadlock
A Primer of Abstract Mathematics, Robert B. Ash
Proofs Without Words, Roger B. Nelsen
Proofs Without Words II, Roger B. Nelsen
A Radical Approach to Real Analysis, David M. Bressoud
She Does Math!, edited by Marla Parker

MAA Service Center
P. O. Box 91112
Washington, DC 20090-1112
1-800-331-1622 fax: 1-301-206-9789

Contents

To the memory of
Wilbur R. Knorr
renowned Archimedes scholar
who died at the age of fifty-one in 1997,
a year before the reappearance of the lost palimpsest

Acknowledgments

Several people have helped to further this book's mission of making Archimedes easily accessible to a wide audience. The readers and the author gratefully wish to thank:

Don Albers, head of publications at the MAA, who responded enthusiastically to my casual suggestion of writing a book on Archimedes, gently reminded me of the project from time to time, and commented on the manuscript.

Anthony Barcellos, who prepared the illustrations and who advised me concerning several expository dilemmas.

My colleagues Don Chakerian and Henry Alder, who read early versions of the manuscript and made substantial suggestions for making the book more readable.

Roland Hoermann, who translated several pages of Heiberg's article from which the extract in Chapter 4 is drawn.

Hope Mayo, Consultant for Manuscripts at Christie's New York auction house, who took me on a guided tour of the Archimedes palimpsest.

Hannah Stein, my wife, who improved innumerable sentences and paragraphs, greatly lightening the task of the reader.

David Traill, the classicist, who advised me on Roman and Greek historiography and number notation.

The anonymous owner of the Archimedes palimpsest for permission to use a photograph of one of its pages.

Introduction

We often hear that Archimedes is one of the three greatest mathematicians of all time. However, when asked, "What did he do?," most of us are at a loss for examples: "Didn't he find the surface area of a sphere?," or "I think he found π," or "Something about running naked out of his bath crying 'Eureka, Eureka.'" I was just as uninformed until I taught the history of calculus and became familiar with Archimedes' many achievements. Each time I gave the course I devoted more time to him until, by the fifth time, 7 of the 20 lectures were devoted to his work, though this unfortunately meant a rather hasty treatment of Newton and Leibniz.

All of Archimedes' surviving works are available in English in the Heath and Dijksterhuis translations, but neither is an easy read. The Heath version follows Archimedes' reasoning so closely that the reader must almost think like an ancient Greek mathematician. On the other hand, the more recent Dijksterhuis version employs an idiosyncratic notation. In either case it is hard to step into the exposition at just any chapter. The reader is almost compelled to become a serious historical scholar to appreciate the arguments.

My goal in writing this little book is to make what I view as Archimedes' most mathematically significant discoveries accessible to the busy people of the mathematical community, whom I think of as anyone who recognizes the equation of a parabola. I hope that high school and college instructors, after reading this book, will invite Archimedes into their classes to enrich their lectures and enlarge their students' historical perspective. The exercises, which serve as a guide and speed bump for the less experienced mathematician, may be skipped without loss of continuity.

Archimedes knew and used far more of the geometry of the conic sections than we ever learn today. This raises an expository problem: How much of that geometry, some quoted from a lost work of Euclid, should I include? If I developed all of it, then this book would grow far longer than I had intended, and Archimedes' major discoveries would disappear in a sea of geometric lemmas.

Practically all of the geometry we need is easily obtained with the aid of affine mappings of the plane. I develop the theory of these mappings from scratch in Appendix A and apply them to the parabola. True, Archimedes did not know about affine mappings, but better a modern proof than none at all. Besides, mappings of the plane appear in the school curriculum, usually serving no purpose. So it is appropriate to demonstrate their power by specific applications.

When approaching each topic, I asked myself, "What mathematics lies at the core of this argument, what timeless gem that transcends the quirks and fashions of the times?" It is that core that I wanted to expose in such a way that the busiest reader could easily follow and appreciate the reasoning. I made no attempt to keep Archimedes' notation or to repeat his explanations word for word. (The reader who wants to look back at the original version may consult Heath's and Dijksterhuis's translations. But keep in mind that these texts, based on earlier translations, may incorporate changes introduced by many editors over the centuries.)

I don't try to cover every topic that Archimedes explored. For instance, I omit his calculation that a mere 10^{63} poppy seeds (hence grains of sand) would "fill the universe" (though, as he says in his introduction, "There are some, King Gelon, who think that the number of grains of sand is infinite.") But I do cover most of his work, certainly enough to demonstrate that his reputation is justly earned.

Of all the remarkable triumphs I describe in the following pages, the one that most impressed me is in *On the Equilibrium of Floating Bodies*, where Archimedes develops the theory of stability of an object floating on water, the basis of naval architecture. (The opening chapters of modern books on the subject have several diagrams similar to those Archimedes drew some 2200 years ago.)

I hope the reader will be as amazed as I by how much Archimedes accomplished with the limited tools at his disposal some 22 centuries ago.

1

The Life of Archimedes?

What we are told about Archimedes is a mix of a few hard facts and many legends. I will describe their sources so that you may decide for yourself how much truth lies in each one. Hard facts—the primary sources—are the axioms of history. Unfortunately, a scarcity of fact creates a vacuum that legends happily fill, and eventually fact and legend blur into each other. The legends resemble a computer virus that leaps from book to book, but are harder, even impossible, to eradicate. Before we try to disentangle them in the case of Archimedes, let's consider a simpler case, one closer to home, which goes back a mere two centuries rather than two millennia. This will put us in the proper mood to deal with the facts and fancies that surround Archimedes.

Little is known of George Washington's boyhood, but that didn't stop Mason Weems, in his popular *Life of George Washington*, published in 1806, from introducing the now-famous incident of the hatchet and the cherry tree:

> "George," said the father, "do you know who killed that beautiful little cherry-tree yonder in the garden?"
>
> "I can't tell a lie, Pa ... I did cut it with my hatchet."
>
> "Run to my arms... ," cried his father... "glad am I that you killed my tree.... "

As early as 1824 Weems was described as the "author of A Washington's Life—not one word of which we believe." In 1887 a biographer of Washington could write, "The material [relating to his boyhood] is rather scanty. The story of the hatchet and the cherry sapling, whether true or not, is singularly characteristic.... Nobody would ever have thought of relating such a story in connection with the boyhood of Napoleon.... "

And we find, in a children's' book published in 1954, "Stories about George Washington as a boy have been retold so often ... that even though we're not sure they really did happen, they have become a part of the story of America. And they do tell us something of the kind of boy he was."

Present-day historians must carefully back up their assertions by citing the evidence or the source. But, as P. G. Walsh, an authority on ancient history, points out in *Livy, His Historical Aims and Methods*, the historians in Archimedes' time did not all abide by such harsh standards: "the majority of histories written in this period [400–200 B.C.] were not so laudable. [The] detailed reasons for such a retrogression ... are undoubtedly connected with the growth of the schools of rhetoric.... [W]riters addicted to 'tragic' techniques sought to thrill their readers by evoking feelings of pity and fear.... Certain types of description are particularly amenable to this type of treatment, such as the fate of conquered cities, or the deaths of famous men."

Clearly we yearn to discover the human side of our heroes, to find out what influences and circumstances shaped them into such unique figures. Perhaps subconsciously we are looking for the magic formula that will turn children into creative adults.

We should be on guard about Archimedes, who invites far more invention than Washington does. There is no material at all about his boyhood and little about his adulthood. However, there is ample evidence that he lived in Syracuse (Syracusa, the city at the southeast corner of Sicily, then part of the Greek world) and was killed there by a Roman soldier in the year 212 B.C., at the end of a siege conducted by the Roman general Marcellus. Legends soon grew up about the manner of his death. Plutarch (about 46–120 A.D.) described it this way:

"... He was alone, examining a diagram closely; and having fixed both his mind and his eyes on the object of his inquiry, he perceived neither the inroad of the Romans nor the taking of the city. Suddenly a soldier came up to him and bade him follow to Marcellus, but he would not go until he had finished the problem.... [T]he soldier became enraged and dispatched him. Others, however, say that the Roman came upon him with drawn sword intending to kill him at once, and that Archimedes ... entreated him to wait ... so that he might not leave the question ... only partly investigated; but the soldier did not understand and slew him. There is a third story, that as he was carrying to Marcellus some of his mathematical instruments, such as sundials, spheres, and angles adjusted to the apparent size of the sun, some soldiers . . . under the impression that he carried treasure in the box, killed him."

Neither Plutarch nor Livy (59 B.C.–17 A.D.), who also wrote about Archimedes, mentions that he told the soldier, "Noli turbare circulos meos." ("Don't disturb my circles.") The first time words were provided for Archimedes was about 30 A.D., when Valerius Maximus had him request, "Noli obsecro istum disturbare." ("Please don't disturb this.") Presumably Archimedes was referring to a diagram he had drawn on the ground or on a table. By the twelfth century, Archimedes is saying, with more feeling and in Greek, "Fellow, stand away from my diagram." I suspect that these variations are pure fiction. Who would have reported them? Would a soldier who had killed Archimedes, against orders from his commanding general, offer this incriminating evidence? Nor do Plutarch and Livy mention his drawing diagrams on the ground. This was a later "improvement."

As for his date of birth we must depend on *The Book of Histories*, a twelfth-century work by the historian Tzetzes, who lived some fourteen centuries after Archimedes. Hardly his contemporary, Tzetzes wrote: "Archimedes the wise, the famous maker of engines, was a Syracusan by race, and worked at geometry till old age, surviving five-and-seventy years." That statement, together with a bit of arithmetic, is the source for asserting that Archimedes was born in 287 B.C.

The tale of Archimedes crying "Eureka, Eureka" goes back to the Roman architect, Vitruvius, who lived in the first century B.C. Archimedes had to find the volume of a sacred wreath (not, as is often said, a crown), allegedly made of pure gold, to determine whether the goldsmith had replaced some gold with silver.

"While Archimedes was turning the problem over, he chanced to come to the place of bathing, and there, as he was sitting down in the tub, he noticed that the amount of water which flowed over the tub was equal to the amount by which his body was immersed. This showed him a means of solving the problem, and he did not delay, but in his joy leapt out of the tub and, rushing naked towards his home, he cried out with a loud voice that he had found what he sought. For as he ran he repeatedly shouted in Greek, 'Eureka, Eureka' (I have found, I have found)."

I doubt that such an elementary observation would have impressed Archimedes enough to stage a celebration, especially when it is compared to his dazzling discoveries about the surface area and volume of a sphere, the center of gravity, and the stability of floating objects. In fact, Archimedes viewed his work on the sphere as his best, as Plutarch reported: "And though he made many elegant discoveries, he is said to have besought his friends and kinsmen to place on his grave ... a cylinder enclosing a sphere, with an inscription giving the proportion by which the volume of the cylinder exceeds that of the sphere."

That must be a hard fact, for Cicero, about 75 B.C., saw the stone: "When I was questor I searched out his tomb, which was shut in on every side and covered with thorns and thickets. The Syracusans did not know of it: they denied it existed at all. I knew some lines of verse which I had been told were inscribed on his gravestone, which asserted that there was a sphere and cylinder on the top of the tomb.

"Now while I was taking a thorough look at everything—there is a great crowd of tombs at the Agrigentine Gate—I noticed a small column projecting a little way from the thickets, on which there was a representation of a sphere and cylinder. I immediately told the Syracusans—their leading men were with me—that I thought that was the very thing I was looking for. A number of men were sent in with sickles and cleared and opened up the place.

"When it had been made accessible, we went up to the base facing us. There could be seen the epitaph with about half missing where the ends of the lines were worn away. So that distinguished Greek city, once also a centre of learning, would have been unaware of the tomb of its cleverest citizen, if it had not learned of it from a man from Arpinum....

"Who in the world is there, who has any dealings with ... culture and learning, who would not rather be this mathematician than that tyrant [Dionysius, who ruled Syracuse for 38 years]? ... [T]he mind of the one was nourished by weighing and exploring theories, along with the pleasure of using one's wits, which is the sweetest food of souls, the other's in murder and unjust acts...."

During his lifetime, Archimedes was famous for his military exploits, in particular for his contribution to the defense of Syracuse against the Romans. The earliest description of this side of Archimedes is in the *Histories* of Polybius (about 204–120 B.C.):

"But in this the Romans did not take into account the abilities of Archimedes; nor calculate on the truth that, in certain circumstances, the genius of one man is more effective than any numbers whatever. However they now learnt it by experience ... Archimedes had constructed catapults to suit every range; and as the ships sailing up were still at a considerable distance, he so wounded the enemy with stones and darts, from the ... longer engines, as to harass and perplex them to the last degree. And when these began to carry over their heads, he used smaller engines graduated according to the range required. Finally, Marcellus was reduced in despair to bringing up his ships under cover of night. But when they had come too near to be hit by the catapults, they found that Archimedes had prepared another contrivance ... "

Polybius goes on to describe other devices Archimedes employed, including grappling hooks that lifted the Roman ships out of the water. With these devices Archimedes forced Marcellus to give up any hope of a successful frontal attack, and inaugurated the role of the warrior scientist, who, especially in this century, has had a far greater impact on the nature of war than have the generals.

The catapults, equipped with ropes and pulleys, exploited the principle of the lever, where distance is traded for force. Plutarch mentions Archimedes' appreciation of the lever: "Archimedes, a kinsman and friend of King Hiero, wrote to him that with a given force it was possible to move any given weight; and emboldened, as it is said, by the strength of the proof, he asserted that, if there were another world and he could go to it, he would move this one."

By the fourth century A.D. this had become, in Pappus' version,the well-known epigram, "Give me a place to stand and I can move the Earth." As we will see in later chapters, Archimedes certainly appreciated the principle of the lever, applying it in his study of area, volume, and center of gravity.

Another legend has it that he set the Roman ships on fire by using mirrors arranged in a parabola to reflect the sunlight on a single burning point. However, there is no mention of this in the three earliest descriptions of the defense of Syracuse, those by Polybius, Livy, and Plutarch. Lucian, writing in the first century A.D., says only that Archimedes used "artificial means." The first mention of mirrors occurs in Galen, around the year 160 A.D., almost four centuries after the death of Archimedes.

The more one thinks of trying to burn ships at a distance in the midst of a battle, the more implausible the project seems: The sun must be bright and at a convenient angle, the boat not bob on the waves, the people holding the mirrors able to focus on the exact same spot while dodging volleys of flying arrows. Moreover, had the mirrors done their work, they would have become a standard weapon; yet there is no sign that they were added to the armaments of the time.

At the end of his study based on the ancient sources and the pertinent physics, D. L. Simms concludes, "The historical evidence for Archimedes' burning mirrors is feeble, contradictory in itself, and the principal and very late authorities for the story are unreliable, while the standard and contemporary authorities are silent.... Modern experiments suggest a burning mirror is highly unlikely to produce ignition on a moving ship."

Where does all this leave us? That Archimedes lived in Syracuse and applied mathematics to the so-called "real world," and died in the assault on his city. Historians conjecture that he may have visited Alexandria, Egypt, which, with its great library, was a major scientific center, and it is possible that he invented the device for raising water for irrigation, the "Archimedes screw." But these assertions are part of that ambiguous and hypothetical world of "perhaps," "maybe," and "it is possible that," where we are left not knowing what to think.

For a more extensive discussion of the sources of information about the life of Archimedes I refer the reader to the opening pages of Dijksterhuis' book and Knorr's supplement at the end.

When all is said and done, we do have many of Archimedes' writings, filtered though they may be through Arab, Latin, and English translations. These, after all, bear the most reliable witness of the man, and are the main reason we are interested in him. The best way to appreciate Archimedes is to follow his mathematical arguments, just as we best appreciate Giotto by viewing his frescoes and Mozart by listening to his music.

2

The Law of the Lever

Once when I was introducing applications to motivate a high school algebra class, I brought in some sticks and weights and asked the students to experiment and find a general rule for finding the positions at which two unequal weights would balance each other. I expected them to discover the law of the lever, namely that the product of one weight by its lever arm equals the product of the other weight by its lever arm. To my surprise and disappointment, they did not discover the rule. Clearly, the law of the lever is not as obvious as I had thought.

Archimedes develops the law as the first step in his theory of the center of gravity. Rather than stating it as a fact, he derives it from more basic principles in the style of Euclid's geometry, written a generation or two before him. Let us see how he does this.

Two weights, W and w, are placed on a horizontal weightless stick. The stick rests on a support, called the fulcrum, as in Figure 1.

Either the stick tilts clockwise or counterclockwise around the fulcrum or it remains horizontal. In this third case, the weights are said to balance each other or to be in equilibrium. Archimedes considers the question, "If W is at a distance D from the fulcrum and w is at a distance d from the fulcrum, what condition on W, w, D, and d corresponds to equilibrium?" The answer, which is part of any introductory physics course, is that the two products, WD and wd, must be equal.

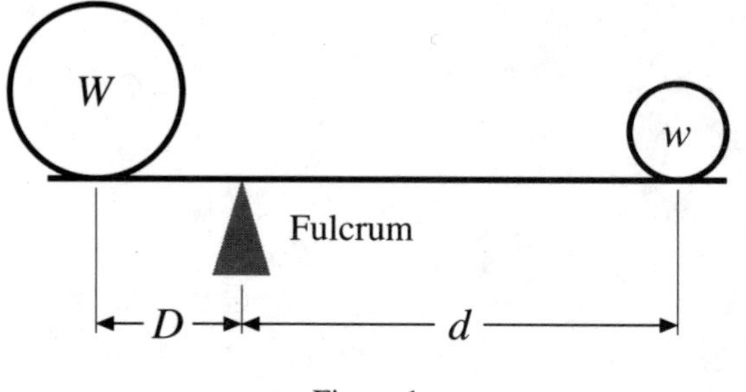

Figure 1

If you look for the law in this form in Archimedes' work you won't find it, for he would have been offended by multiplying two such different quantities as a weight and a length. Instead, he expresses balance by the equality of two proportions: $W : d = w : D$. (W is to d as w is to D.) He also states that the weights balance at distances inversely proportional to their magnitudes.

He distinguishes the two cases that we would describe as "W/w rational" and "W/w irrational." In his words they would be "W and w commensurable" and "W and w incommensurable." (Two numbers, W and w, are commensurable when there is a third number, m, such that $W = pm$ and $w = qm$, where p and q are whole numbers. In other words, W and w have a common measure, m. If there isn't such a common measure, then W and w are said to be incommensurable.)

Exercise 1. Show that W/w is rational only when W and w are commensurable.

Here are the assumptions with which Archimedes begins his theory of weights on a balance:

Assumption 1. Equal weights at equal distances from the fulcrum balance. Equal weights at unequal distance from the fulcrum do not balance, but the weight at the greater distance will tilt its end of the lever down.

Assumption 2. If, when two weights balance, we add something to one of the weights, they no longer balance. The side holding the weight we increased goes down.

Assumption 3. If, when two weights balance, we take something away from one, they no longer balance. The side holding the weight we did not change goes down.

Now let's see how Archimedes applies these assumptions.

Proposition 1. Weights that balance at equal distances from the fulcrum are equal.

Proof. If they are not equal, remove from the greater weight the difference of the two weights. We now have two equal weights at equal distances from the fulcrum. According to Assumption 3, they do not balance. This contradicts Assumption 1.

Proposition 2. Unequal weights at equal distances from the fulcrum do not balance, but the side holding the heavier weight goes down.

Proof. Take away from the heavier weight the difference between it and the lighter weight. By Assumption 1 the remaining two weights, being equal, balance. If we put back the weight we removed, Assumption 2 asserts that the weights now do not balance and that the heavier weight tilts the lever down at its side.

In proving the next proposition, Archimedes tacitly assumes that two weights do have a balancing point and that it lies somewhere between them.

Proposition 3. Unequal weights balance at unequal distances from the fulcrum, the heavier weight being at the shorter distance.

Proof. Say that the heavier weight is W, placed at A, and the lighter weight, w, is at B and that they balance about the fulcrum C, as in Figure 2.

Remove $W - w$ from the heavier weight W, leaving two equal weights. By Assumption 3, the remaining weights do not balance, but w goes down. But this can't happen for the following reasons.

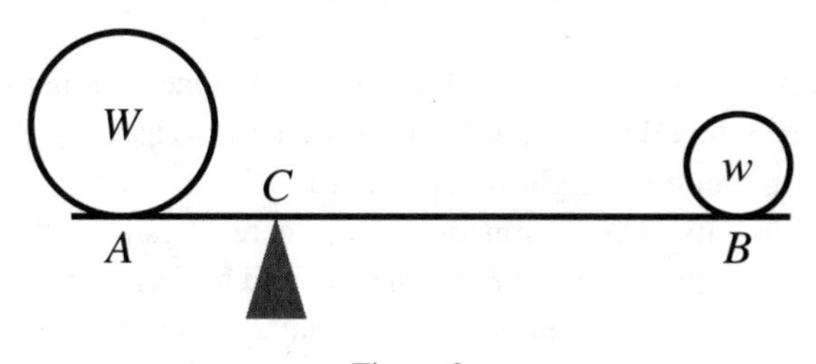

Figure 2

Either AC equals CB or it is greater than CB or it is less than CB. Archimedes rules out the first two possibilities.

If AC equals CB, by Assumption 1, the remaining weights balance. (I use the symbol XY to denote the line segment with ends X and Y and also its length.) If AC is greater than CB, then again by Assumption 1 the weight at A will go down. Thus AC is less than CB.

The next proposition is the key to Archimedes' theory of the lever.

Proposition 4. If two equal weights have different centers of gravity, then the center of gravity of the two together is the midpoint of the line segment joining their centers of gravity.

Before we look at Archimedes' brief proof, let us examine this statement closely. Though the notion of "center of gravity" appears in this proposition (and in much of Archimedes' work), nowhere does he define the phrase. He mentions another of his works on gravity, *Equilibrium*, which has not survived, but I doubt that "center of gravity" is defined there either. Later Greek mathematicians defined it in physical terms, based on experiment, for instance as the point through which all the balancing lines pass. (If a flat object placed horizontally on a line rests in that position, the line is called a "balancing line.") A useful mathematical definition depends on concepts from integral calculus. The typical calculus book first defines a balancing line by requiring that a certain integral equals zero. Then it defines the center of gravity as the intersection of two perpendicular balancing lines. Of course, the text should show that no matter which pair of perpendicular balancing lines is used, their intersection remains the same. That textbooks usually skip this important step suggests that the definition of the center of gravity is not a trivial matter. In any case we see why Archimedes doesn't define it: he simply lacks the necessary mathematical machinery. Instead, as we will see in the next chapter, he approaches the concept axiomatically.

How then are we to interpret Proposition 4? One way is to picture each weight as concentrated in a single point. Another is to assume in the problems we treat that the entire weight acts as if it is all located at one point, called its center of gravity. The confusion here, where physics and the axiomatic method of pure mathematics blur together, should be forgiven. After all, we are witnessing the very first attempt to apply mathematics to achieve an orderly development of a physical theory, in this case, statics.

Now to Archimedes' proof of Proposition 4.

Proof. Let the equal weights be w and w, with centers of gravity located at A and B, and let M be the midpoint of the segment AB. Assume that the weights balance at C, a point different from M, as in Figure 3.

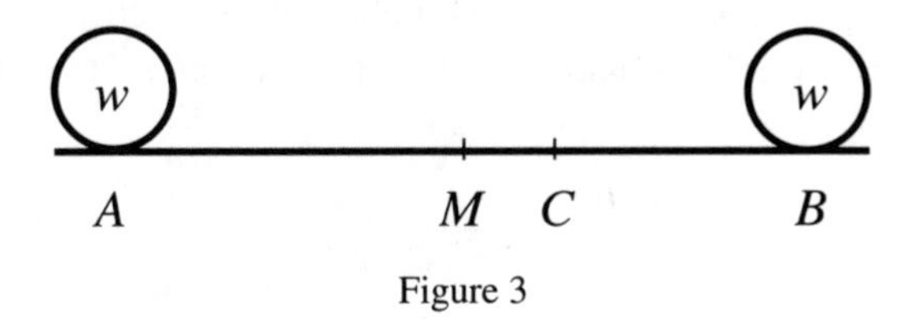

Figure 3

The distance AC is not equal to the distance CB. Therefore, by Assumption 1, the two weights do not balance around C, no matter how C is chosen, as long as it is different from M. This contradiction implies that M must be the balancing point.

Note that in the preceding proof Archimedes tacitly assumes that any two weights have a center of gravity, that is, a balancing point. Proposition 4 is essentially a restatement of Assumption 1 in terms of center of gravity.

Surprisingly, the following corollary is the key to Archimedes' development of the law of the lever.

Corollary. If an even number of equal weights have their centers of gravity situated along a straight line such that the distances between consecutive weights are all equal, then the center of gravity of the entire system is the midpoint of the line segment joining the centers of gravity of the two weights at the middle.

Figure 4 illustrates the corollary. The center of gravity of the eight weights is at C.

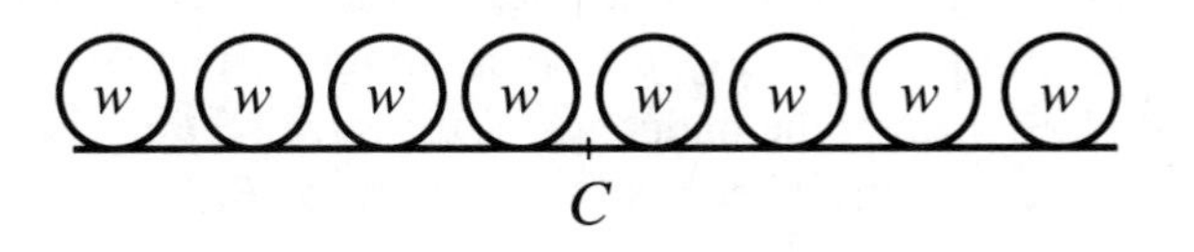

Figure 4

The corollary follows from Proposition 4: group the weights in pairs that have the midpoint C as their center of gravity.

The Law of the Lever

Now Archimedes is ready to develop the law of the lever. First he considers the case where the two weights are commensurable.

Proposition 5. Commensurable weights balance at distances from the fulcrum that are inversely proportional to the magnitudes of the weights. More precisely, if commensurable weights W and w are at distances D and d from the fulcrum, then

$$\frac{D}{d} = \frac{1/W}{1/w} = \frac{w}{W}.$$

Proof. For convenience we take a specific case, where the ratio of the weights is, say, 5 to 3, that is, $W/w = 5/3$. Let W and w be located at A and B respectively, as in Figure 5.

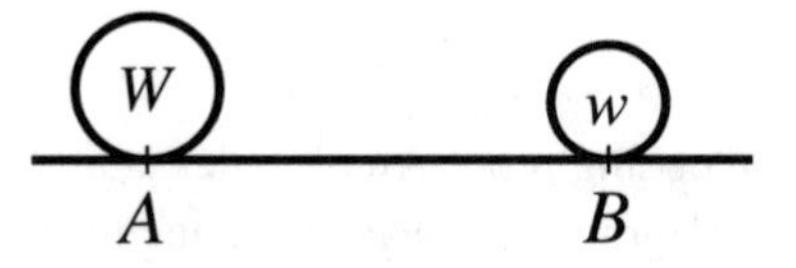

Figure 5

Let M be their balancing point. We wish to show that AM is to BM as $\frac{1}{5}$ is to $\frac{1}{3}$, that is,

$$\frac{AM}{BM} = \frac{1/5}{1/3} = \frac{3}{5}.$$

In short, we wish to show that the ratio between AM and BM is 3 to 5.

Cut the segment AB into $5 + 3 = 8$ equal sections and divide the weight W into $2 \times 5 = 10$ equal weights. Place five of them at the midpoints of the five sections just to the right of A, one in each section, and five of them in congruent sections to the left of A. Similarly, divide the weight w into $2 \times 3 = 6$ equal weights and place them at the midpoints of the three sections just to the left of B and in three congruent sections to the right of B. Figure 6 shows the arrangement.

By the preceding proposition, the collection of 10 weights, each $W/10$, has the same center of gravity as W does, namely A. Similarly, the six weights, each $w/6$, has B as its center of gravity. Archimedes now has a system of 16 equal weights (they are equal since $W/10 = w/6$) equally spaced.

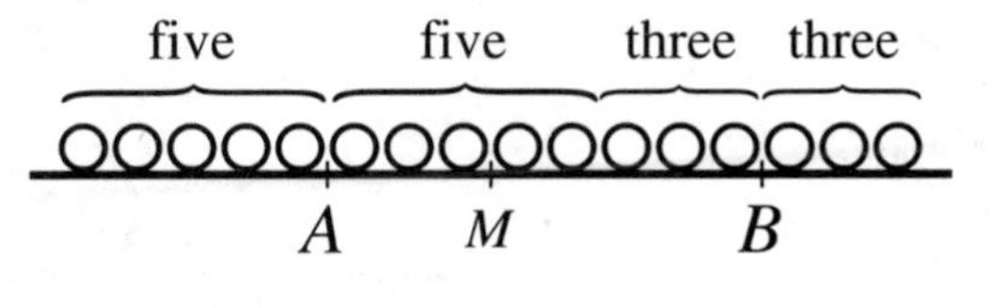

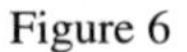
Figure 6

By the corollary to Proposition 4, the collection of 16 weights balances around the midpoint M of the segment holding the 16 weights. This implies that the ratio of the lever arms of W and w, namely AM and BM, is 3 to 5, since AM has three of the little weights and BM has five of them. This is precisely what Archimedes claims.

Exercise 2. Prove Proposition 5 in general, when $W/w = m/n$, where m and n are positive integers.

Archimedes then deduces the incommensurable case from the preceding commensurable case. In his argument he uses an assertion that is almost a consequence of Proposition 5. Let us look at it in advance.

Say that W/w is rational and $W \cdot D$ is less than $w \cdot d$. If weight W is placed at a distance D from the fulcrum and weight w at a distance d on the other side, what will happen? To answer, introduce D' such that $W \cdot D' = w \cdot d$. Note that D' is larger than D. Then by Proposition 5, W at a distance D' balances w at a distance d. It seems plausible that shrinking the distance from W to the fulcrum from D' to D will force W to rise. Archimedes assumes this in the following proof, though it isn't a logical consequence of his assumptions.

Proposition 6. Incommensurable weights balance at distances inversely proportional to their magnitudes.

Proof. Let the weights be W and w at respective distances D and d from the fulcrum. Assume that $WD = wd$ and that the two weights do not balance. To be specific, assume that the weight W goes down, as indicated in Figure 7.

Remove a small amount from the weight W to obtain a weight W' such that W' still goes down but W' and w are commensurable. However, by the plausible observation we made after Proposition 5, since $W'D$ is less than wd, W' rises. This contradiction—that W' both goes down and rises—proves the proposition.

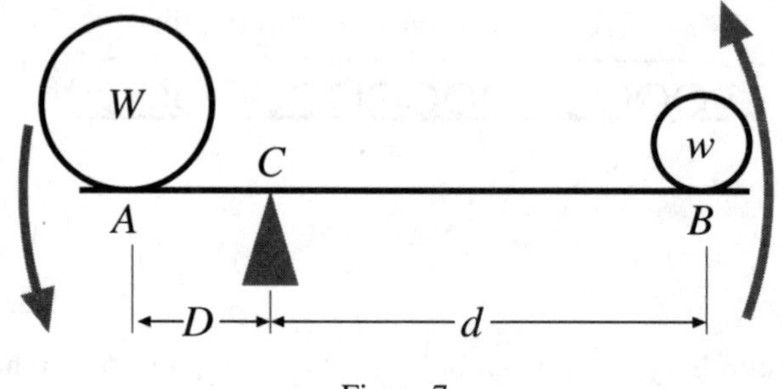

Figure 7

Strictly speaking, the preceding proof depends not only on the earlier assumptions and the plausible observation but also on physical intuition. How do we know that there is a W' such that W'/w is rational and W' is still too heavy? To be completely rigorous, Archimedes would have had to add some more assumptions.

In any case, Archimedes is now free to use the law of the lever when dealing with centers of gravity. In particular, he identifies the center of gravity of a pair of weights as the point (fulcrum) around which the two weights balance. In the next chapter he uses it to find the centers of gravity of several figures. Then, as we will see, he employs centers of gravity to find areas and volumes and to analyze the equilibrium of a floating object.

3

Center of Gravity

We can think of the law of the lever developed in the preceding chapter as describing the center of gravity of two objects, if we know the center of gravity of each one. The point on which they balance, the fulcrum, is their center of gravity. Although Archimedes never defines the term "center of gravity," he uses this principle to deal with the centers of gravity of solids and of flat objects such as triangles and sections of parabolas.

Archimedes writes about the center of gravity as though it were already a common concept in the Greek scientific community, where it was called $\kappa\varepsilon\nu\tau\rho o\nu\ \tau o\iota\ \beta\alpha\rho\varepsilon o\sigma$ (kentron toi bareos). We get our word 'center' from 'kentron' (sharp point) and 'barometer' from 'bareos' (weight), so their phrase is practically identical with ours.

Later Greek authors offered three definitions of the center of gravity, which were probably already familiar to Archimedes. Note that these definitions are physical rather than mathematical. For convenience, consider the objects to be flat, not solid.

The three definitions of the center of gravity are as follows:

1. The point at which the object can be suspended and remain motionless, no matter in which position it is placed. (See Figure 1, where the center of gravity is labeled C.)

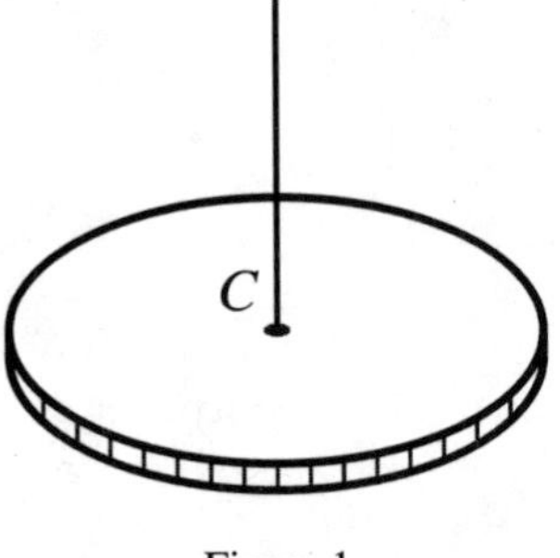

Figure 1

2. The point common to all the vertical lines through points of suspension. (See Figure 2.)

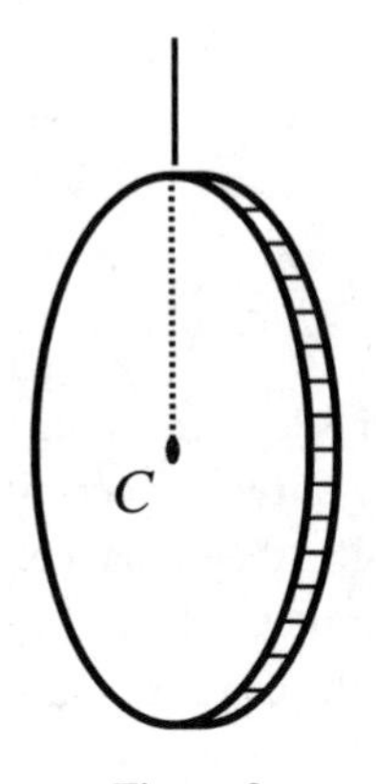

Figure 2

3. The point common to all the lines on which the object balances. (See Figure 3.)

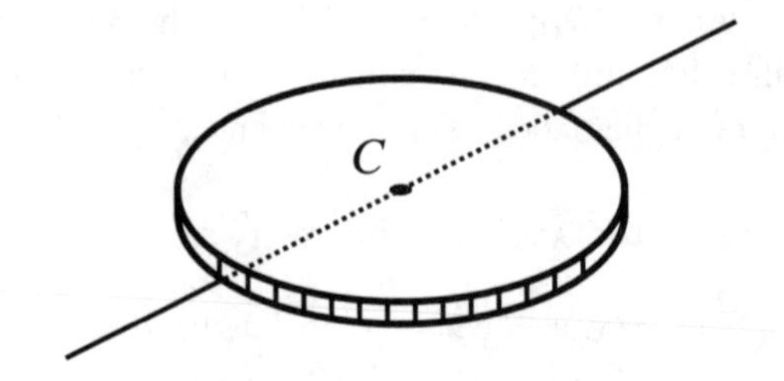

Figure 3

None of these definitions could serve Archimedes in his geometric reasoning. Even if he assumed, say, that there is a point through which the balancing lines pass, the assumption would not help him find that point mathematically. Instead, he needs properties of the center of gravity that enable him to compute that point theoretically. This he does by stating three principles concerning the center of gravity and using them to calculate centers of gravity. It is understood that the objects have a uniform density. The first two principles are fresh axioms or assumptions. The third is a restatement of the law of the lever, extended to whole areas.

Before we state the principles, we define the term "convex set," which appears in the second principle. A set in the plane is convex if it has no indentations. More precisely, if P and Q are any points in the set, then the entire line segment joining them also lies in the set, as illustrated in Figure 4. Triangles, circles, and parallelograms are examples of convex sets.

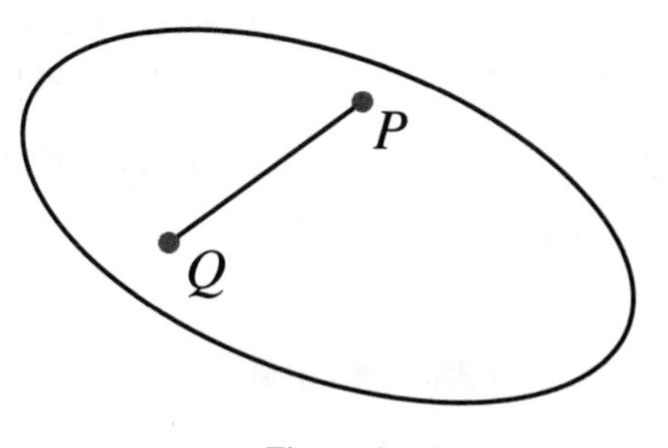

Figure 4

The three principles are:

I. The centers of gravity of congruent or similar figures are similarly situated.

II. The center of gravity of a convex figure lies within the figure.

III. If an object is cut into two pieces, its center of gravity, C, lies on the line segment joining the centers of gravity of the pieces. Moreover, if the pieces are R and S, with respective centers of gravity A and B, then

$$CA \times \text{area of } R = CB \times \text{area of } S.$$

Figure 5 illustrates the relation between these three centers of gravity.

This third principle is suggested by the principle of the lever developed in the preceding chapter.

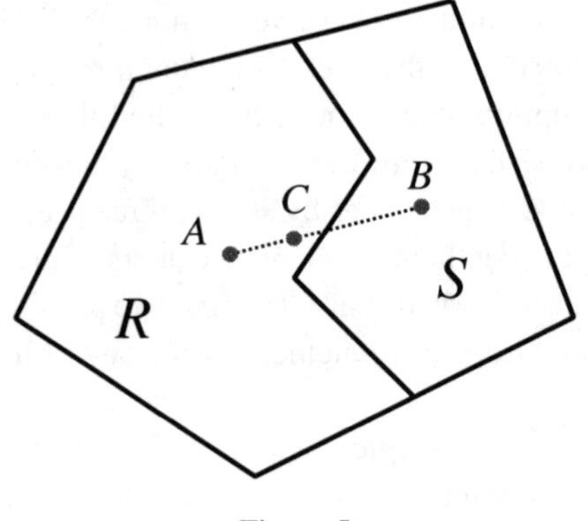

Figure 5

Note that when dealing with equilibrium, Archimedes assumes that the region can be replaced by its center of gravity. Observe also that the center of gravity of the whole region lies between the centers of gravity of its two parts. Archimedes uses this fact in his study of the equilibrium of floating objects (see Chapter 8).

Now let's watch as he uses these three principles in his treatment of the center of gravity of a parallelogram.

The Center of Gravity of a Parallelogram

Let $ABCD$ be a parallelogram and E and F the midpoints of the opposite sides AD and BC, respectively, as in Figure 6. Let H be the center of gravity of the

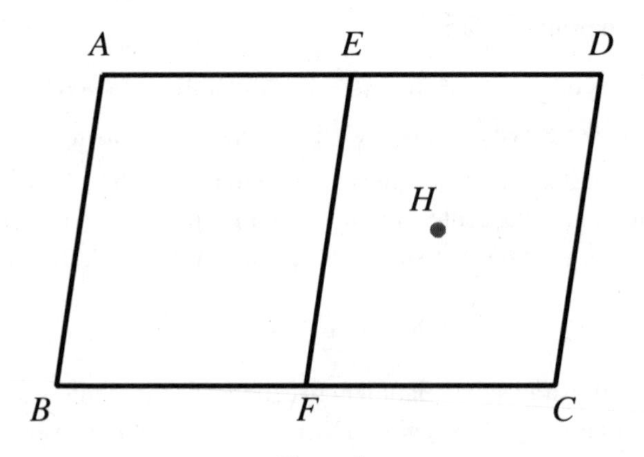

Figure 6

parallelogram. Either H is on the line joining E and F, or H is not on that line. Archimedes assumes the second case and draws a contradiction. Since that rules out the second possibility, it follows that H must lie on the line joining E and F. Let us see how Archimedes reaches a contradiction.

By his second principle, H lies somewhere in the parallelogram, as shown in Figure 6.

The line through H and parallel to AD meets EF at a point K, as in Figure 7.

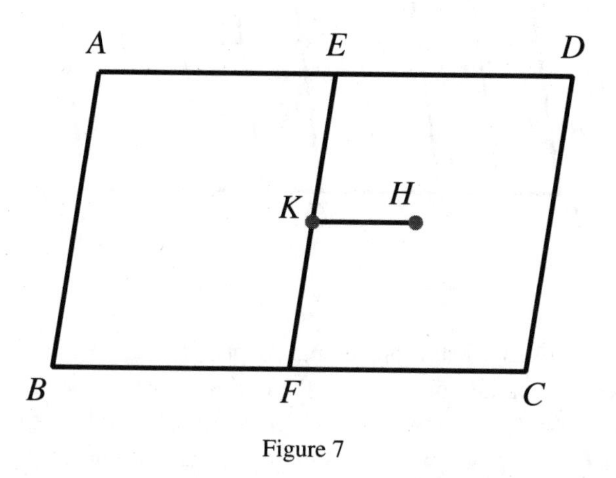

Figure 7

Divide ED and AE into equal pieces shorter than KH, and draw lines parallel to AB through the points of division, as in Figure 8.

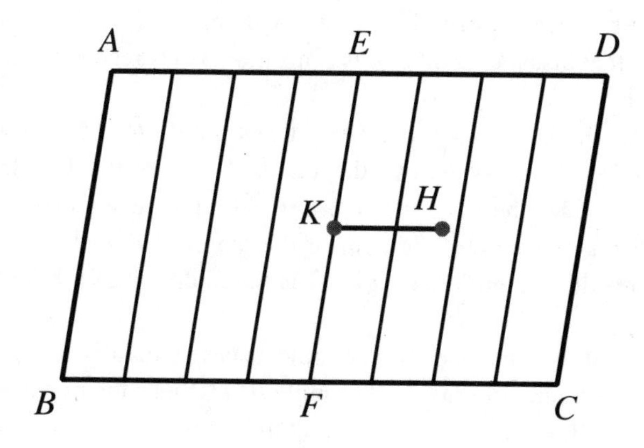

Figure 8

The original parallelogram is now cut into an even number of smaller, congruent parallelograms. By the first two principles, their centers of gravity lie within them and are similarly placed. In Figure 9 they are shown as dots.

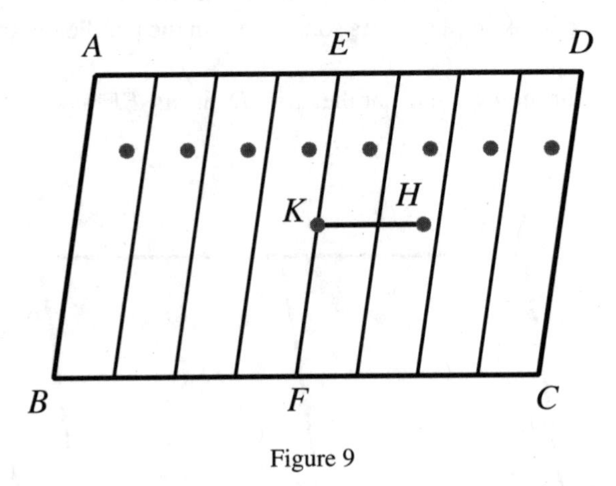

Figure 9

By the corollary in the preceding chapter, the center of gravity of the collection of little parallelograms is at the midpoint of the line segment joining the centers of gravity of the two parallelograms in the middle, the ones adjacent to the line *EF*. This point, which lies in one of the two middle parallelograms or on their common border, cannot be *H*. This contradiction implies that *H* lies on *EF*.

A similar argument shows that the center of gravity of the parallelogram *ABCD* also lies on the line joining the midpoints of the other two sides, *AB* and *CD*. Therefore the center of gravity of a parallelogram is at the intersection of the two lines that bisect opposite sides. This point is also the intersection of the two diagonals.

Archimedes also gives a direct proof that the center of gravity of a parallelogram lies on each of its two diagonals. Again let the parallelogram be *ABCD* and consider the diagonal *BD*. Let *F* be the center of gravity of the triangle *ABD*. Let *G* be the midpoint of the diagonal *BD*. Reflect *F* through *G*. Call the resulting point *H*. That is, *G* is the midpoint of the line *FH*, as in Figure 10.

By the assumption that congruent objects have similarly placed centers of gravity, *H* is the center of gravity of triangle *BCD*. Since the two triangles have equal areas, the third principle asserts that the center of gravity of their union, the parallelogram, has its center at the midpoint of *FH*, namely, *G*. Since *G*

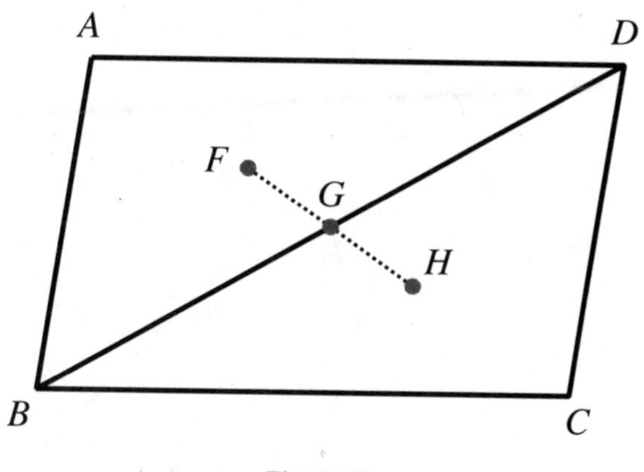

Figure 10

also lies on the other diagonal, the center of gravity of the parallelogram lies on both diagonals.

Exercise 1. Show by geometry that the diagonal AC passes through G.

The Center of Gravity of a Triangle

Where is the center of gravity of a triangle? If the triangle were equilateral, Archimedes' first principle about congruent figures would be enough to answer the question. One may apply a rotation by 120 degrees to rotate such a triangle into itself. Since only the center of rotation is left fixed and the center of gravity is taken to itself, it must be the center of the rotation. That point can be described as the intersection of the lines joining each vertex to the midpoint of the opposite side.

Archimedes gives two proofs that the center of gravity of any triangle lies on each line that joins a vertex to the midpoint of the opposite side, a so-called "median." I will present one of his two proofs.

Let ABC be a triangle and D the midpoint of side BC, as in Figure 11.

Assume that the center of gravity of the triangle, H, is not on AD. To be specific, assume that it lies to the right of the median AD. Let E be the midpoint of AC and let F be the midpoint of AB. Using D, E, and F, cut the triangle into two triangles and a parallelogram, as in Figure 12.

The two smaller triangles, being similar to triangle ABC, have their centers of gravity similarly located. Let K be the one for triangle BFD and L the one

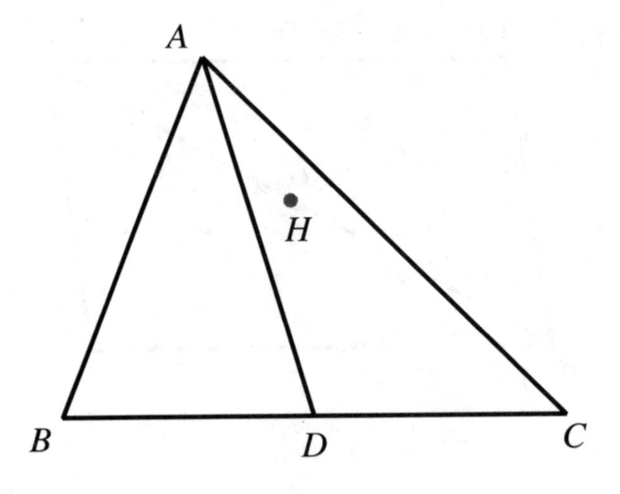

Figure 11

for triangle *DEC*. Let *M* be the center of gravity of the parallelogram *AEDF*. As was just shown, it lies on the intersection *EF* and *AD*, as in Figure 13.

Note that *AH*, *FK*, and *EL* are parallel and that *FK* has the same length as *EL*, namely half the length of *AH*. Let *N* be the midpoint of *KL*, as shown

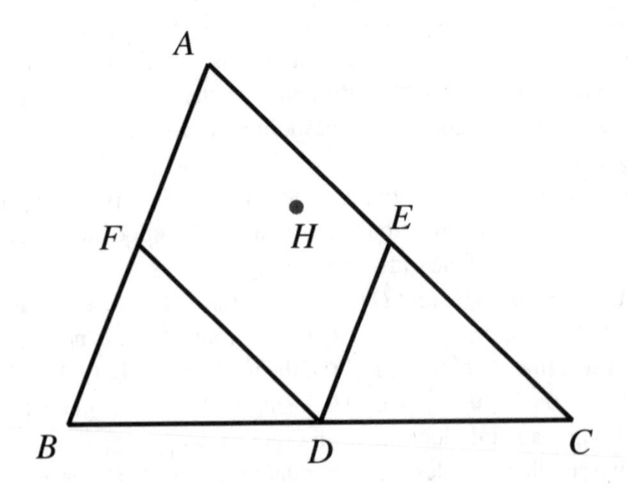

Figure 12

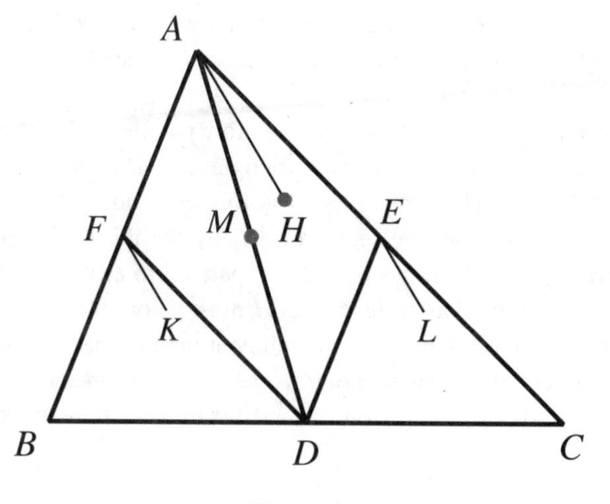

Figure 13

in Figure 14. It is the center of gravity of the two triangles *BFD* and *DEC*, considered together as one figure.

In Figure 14, *MN* looks parallel to *EL* (and *FK*). Let's check that it is.

First of all, *M* is the midpoint of *FE*, a diagonal of parallelogram *AFDE*. Now, since *FK* and *EL* are parallel and of equal length, *FKLE* is a parallelogram. Thus *FE* and *KL* have the same length. Hence *FM*, which is half of *FE*, equals

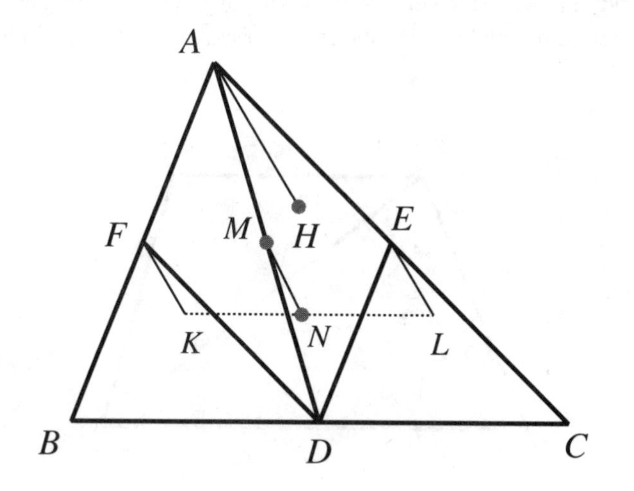

Figure 14

half of KL, which is KN. Thus, since FM and KN are parallel and of equal length, $FKNM$ is a parallelogram. Therefore, as suspected, MN is parallel to FK.

MN is parallel to EL (and FK) and therefore parallel to AH. N is the center of gravity of the union of the two small triangles and M is the center of gravity of the parallelogram. The center of gravity of their union, which is the whole triangle ABC, lies on the line segment MN, by the third principle. But that center of gravity, H, which lies on the line parallel to MN through A, cannot lie on MN. This contradiction shows that H must lie on AD.

Note that the preceding physical argument implies that the three medians of a triangle intersect at a single point. A geometric argument shows that this point lies two thirds of the way from each vertex to the midpoint of the opposite side.

Exercise 2. Using Archimedes' three principles, how would you find the center of gravity of a convex quadrilateral?

Using his three principles, Archimedes also determines the center of gravity of a trapezoid. I will just sketch his method and mention a major problem it raises.

Let the trapezoid be $ABCD$, with AD parallel to BC, as in Figure 15. Archimedes chooses one of its two diagonals and uses it to divide the trapezoid into two triangles, say the diagonal AC.

Already knowing the centers of gravity of triangles ABC and ADC, he applies his third principle to find the center of gravity of the trapezoid.

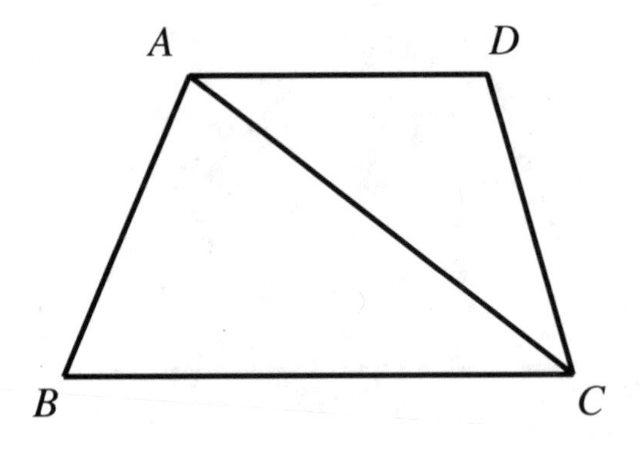

Figure 15

Exercise 3. Assume that the lines on AB and CD meet at E. By viewing the trapezoid in Figure 15 as the difference of two triangles, BCE and ADE, show that its center of gravity lies on the line joining the midpoints of its parallel sides. (This step is part of Archimedes' treatment of the trapezoid.)

There is, however, something serious to worry about. What if Archimedes chooses the other diagonal, BD? He then must carry out calculations using a completely different pair of triangles. How can he be sure that he will obtain the same point as before for the center of gravity of the trapezoid?

Of course, physical experiments convince us that an object has a unique center of gravity. However, we're not sure that the three principles adequately imitate ("model") the physics. Archimedes may not have worried about this concern. However, Olaf Schmidt in 1975 did worry about it. In an article titled "A System of Axioms for the Archimedean Theory of Equilibrium and Centre of Gravity," in the journal *Centaurus*, he showed that the three principles, modified only a little, do provide a mathematical basis for the theory of centers of gravity for polygons.

But that may not be the last word on the matter. If you review Archimedes' first argument for the parallelogram, you will notice that he does not use the full power of his third principle. He needs only the first half of this principle, namely that the center of gravity of the union of two pieces lies somewhere on the line joining their centers of gravity. This raises a question: Could the third principle be weakened to only its first part, and still, together with Principles I and II, provide an adequate description of the physics?

Exercise 4. Do Principles I and II and the first part of Principle III suffice to determine the center of gravity of a polygon? (I don't know.)

Archimedes felt sufficiently at ease with his theory of centers of gravity to apply it to geometric questions. The next two chapters are devoted to this application, which he calls his "mechanical method." Then, in Chapter 8 we will see that he bases his study of floating bodies on properties of centers of gravity.

4

Big Literary Find in Constantinople

Later authors referred to a work of Archimedes called "The Method." Scholars up through the 19th century considered this work lost forever, but in 1906 J. L. Heiberg, the Danish mathematical historian, discovered a thousand-year-old manuscript in Constantinople (now known as Istanbul) that contained the lost work. It was a parchment palimpsest, which is defined as "a document, written upon several times, with remnants of earlier, imperfectly erased writing still visible." Archimedes' work was only lightly washed out and its lines, running at right angles to newer writing that described a ritual of the Greek Orthodox Church, could still be read.

News of this remarkable find appeared in the New York Times in July 1907. I include a facsimile of the front page report in order to convey the significance of the find and a bit of the excitement it generated almost a century ago.

An editorial the next day, headlined "The Archimedean Find," began

> The announcement that a Danish philologist has found and copied a hitherto unsuspected work by Archimedes in a convent at Constantinople will be received with varied emotions. Mathematicians will look forward with eagerness to an unpublished work of one whom they look back upon as the most ancient and illustrious of their guild. Schoolboys will be filled with prescient and anticipatory shudderings, as apprehensive of yet another mathematical textbook.

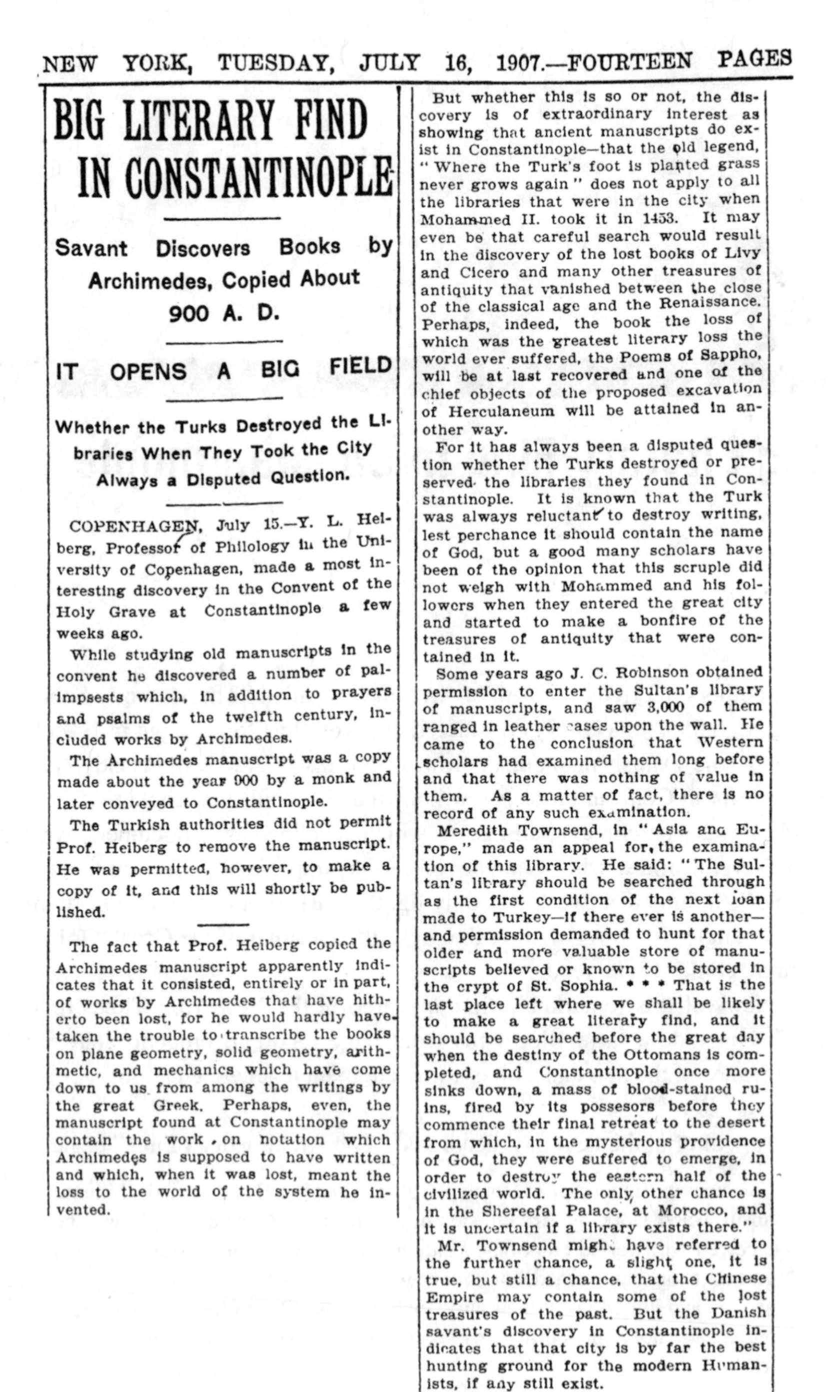

NEW YORK, TUESDAY, JULY 16, 1907.—FOURTEEN PAGES

BIG LITERARY FIND IN CONSTANTINOPLE

Savant Discovers Books by Archimedes, Copied About 900 A. D.

IT OPENS A BIG FIELD

Whether the Turks Destroyed the Libraries When They Took the City Always a Disputed Question.

COPENHAGEN, July 15.—Y. L. Heiberg, Professor of Philology in the University of Copenhagen, made a most interesting discovery in the Convent of the Holy Grave at Constantinople a few weeks ago.

While studying old manuscripts in the convent he discovered a number of palimpsests which, in addition to prayers and psalms of the twelfth century, included works by Archimedes.

The Archimedes manuscript was a copy made about the year 900 by a monk and later conveyed to Constantinople.

The Turkish authorities did not permit Prof. Heiberg to remove the manuscript. He was permitted, however, to make a copy of it, and this will shortly be published.

The fact that Prof. Heiberg copied the Archimedes manuscript apparently indicates that it consisted, entirely or in part, of works by Archimedes that have hitherto been lost, for he would hardly have taken the trouble to transcribe the books on plane geometry, solid geometry, arithmetic, and mechanics which have come down to us from among the writings by the great Greek. Perhaps, even, the manuscript found at Constantinople may contain the work on notation which Archimedes is supposed to have written and which, when it was lost, meant the loss to the world of the system he invented.

But whether this is so or not, the discovery is of extraordinary interest as showing that ancient manuscripts do exist in Constantinople—that the old legend, "Where the Turk's foot is planted grass never grows again" does not apply to all the libraries that were in the city when Mohammed II. took it in 1453. It may even be that careful search would result in the discovery of the lost books of Livy and Cicero and many other treasures of antiquity that vanished between the close of the classical age and the Renaissance. Perhaps, indeed, the book the loss of which was the greatest literary loss the world ever suffered, the Poems of Sappho, will be at last recovered and one of the chief objects of the proposed excavation of Herculaneum will be attained in another way.

For it has always been a disputed question whether the Turks destroyed or preserved the libraries they found in Constantinople. It is known that the Turk was always reluctant to destroy writing, lest perchance it should contain the name of God, but a good many scholars have been of the opinion that this scruple did not weigh with Mohammed and his followers when they entered the great city and started to make a bonfire of the treasures of antiquity that were contained in it.

Some years ago J. C. Robinson obtained permission to enter the Sultan's library of manuscripts, and saw 3,000 of them ranged in leather cases upon the wall. He came to the conclusion that Western scholars had examined them long before and that there was nothing of value in them. As a matter of fact, there is no record of any such examination.

Meredith Townsend, in "Asia and Europe," made an appeal for the examination of this library. He said: "The Sultan's library should be searched through as the first condition of the next loan made to Turkey—if there ever is another—and permission demanded to hunt for that older and more valuable store of manuscripts believed or known to be stored in the crypt of St. Sophia. * * * That is the last place left where we shall be likely to make a great literary find, and it should be searched before the great day when the destiny of the Ottomans is completed, and Constantinople once more sinks down, a mass of blood-stained ruins, fired by its possesors before they commence their final retreat to the desert from which, in the mysterious providence of God, they were suffered to emerge, in order to destroy the eastern half of the civilized world. The only other chance is in the Shereefal Palace, at Morocco, and it is uncertain if a library exists there."

Mr. Townsend might have referred to the further chance, a slight one, it is true, but still a chance, that the Chinese Empire may contain some of the lost treasures of the past. But the Danish savant's discovery in Constantinople indicates that that city is by far the best hunting ground for the modern Humanists, if any still exist.

The first prediction was right: the discovery helped mathematical historians fill a big gap in our understanding of Archimedes. The second was wrong, for the method did not join the standard curriculum of algebra and geometry.

Since newspapers notoriously do not get all the facts right, I also include an excerpt from Heiberg's own report, published, in German, in the journal Hermes in 1907:

> In connection with the revision of my Archimedes edition (originally published more than 25 years ago) I was made aware by Professor Schöne of the fact that Papadopolous Kerameus' fourth volume of *The Jerusalem Library* that appeared in 1899 listed a palimpsest containing mathematical material. Fortunately he included several lines of the underneath text that sufficed to demonstrate that it was a work of Archimedes. Inasmuch as an attempt through diplomatic channels failed to bring the manuscript to Copenhagen, I went during the summer recess of 1906 to Constantinople, where the manuscript resides in the library of the monastery. As a result of the gracious friendliness of the librarian Mr. Tsoukaladakis, I was able within a short time to compare and copy a large portion of the manuscript. Since it soon became evident that the manuscript also contained new material that, being quite indecipherable without thorough study, would require much more time than I had at my disposal, I had the relevant pages as far as possible photographed.

Heiberg included a photo of a page from the manuscript, which appears on page 30. The page reproduced on the cover, made almost a century later, is digitally enhanced to emphasize the original writing. It is provided courtesy of Christie's Images Ltd. 1999. Both are from Archimedes' work *On Floating Bodies*.

Sometime in the 1920s the manuscript vanished again. Then, some 70 years later, it surfaced, as this headline in the New York Times of October 27, 1998 announced:

Ancient Archimedes Text Turns Up, and It's for Sale

The article began

> A millennium-old volume, defaced by mildew and scorched around the edges, will go on the auction block on Thursday, and scholars hope that expert examination of the relic will reveal nuances in the thinking of Archimedes — the greatest mathematical genius of the ancient world.

The manuscript, which had been in the possession of an anonymous French family, was to be auctioned October 29 by Christie's in New York, which

Hermes XLII. Bd. 1907. S. 235.

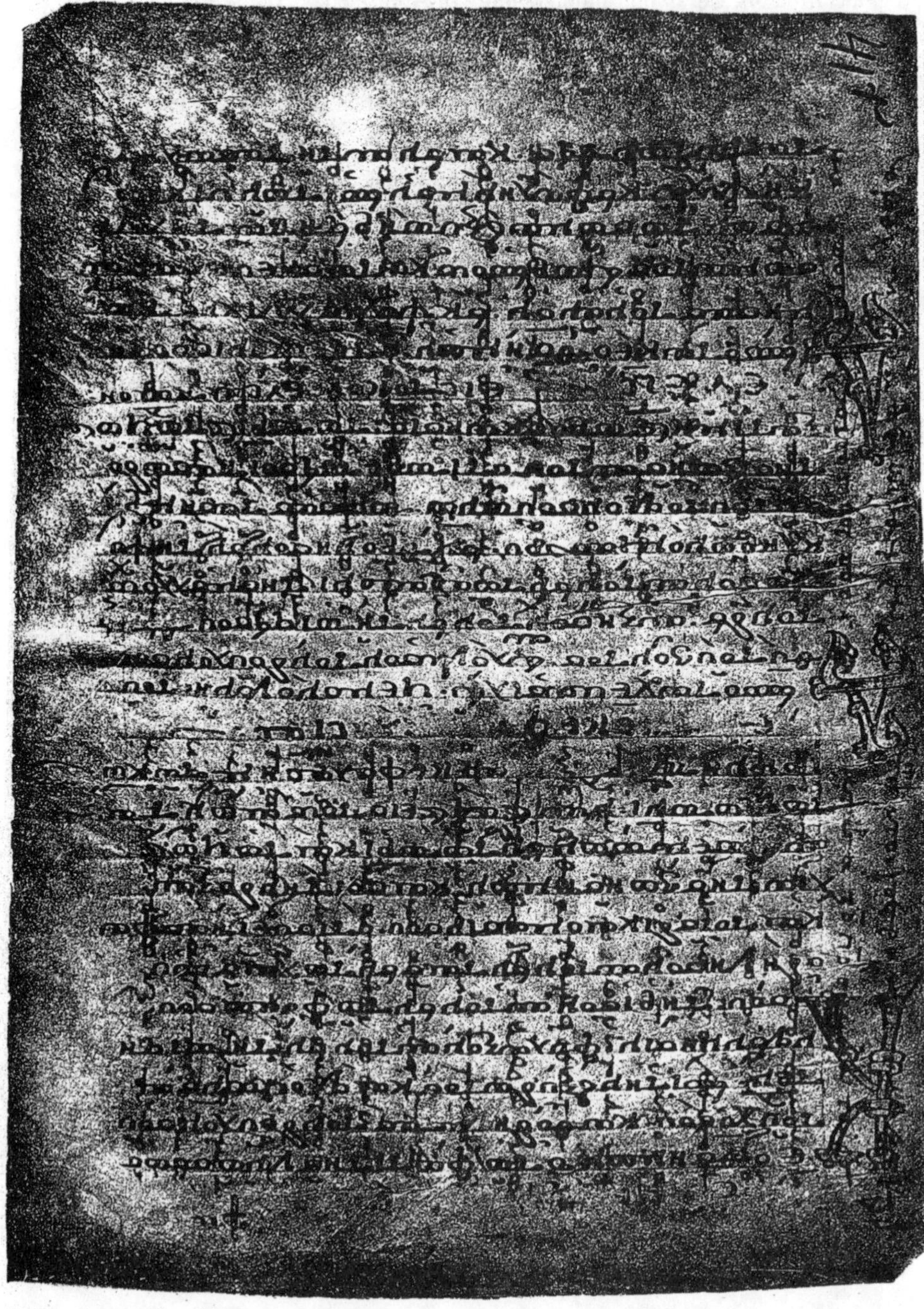

Lichtdruck von Albert Frisch, Berlin W 35

expected it to sell for $800,000 to $1,200,000. The Greek Orthodox Patriarchate of Jerusalem brought action in Federal Court the day before the sale to block the auction. However, the judge, citing French law that asserts that a party who buys an object and owns it for at least thirty years acquires full title, denied the request.

The Greek counsel bid but dropped out at $1,900,000. The winning bid was $2,000,000 (plus a 10% commission to the auction house) from an anonymous American collector, who said that scholars will have access to the manuscript. So the manuscript passed from one anonymous owner to another. However, the Patriarchate threatened to take legal action to recover the palimpsest, which they asserted was stolen. Even so, the new owner permitted it to be put on display to the general public for the first time in a thousand years, as the New York Times of February 19, 1999 reported.

Eureka!

When the oldest surviving copy of the important mathematical works of Archimedes sold at Christie's in New York this fall for $2.2 million, the buyer was a low-profile American collector who wished to remain anonymous. So mathematicians, historians, scientists and scholars thought they would never see the Archimedes palimpsest again.

But this week the Walters Art Gallery in Baltimore announced that it would show the manuscript in a special exhibition from June 20 through Sept. 5.

Gary Vikan, director of the Walters, had gotten in touch with Simon Finch, the London dealer who bought the manuscript on behalf of the collector. It turned out that the new owner was a supporter of the Walters.

Mr. Vikan, who is a Byzantine scholar, said the Walters was the right place to show the work by Archimedes, the Greek mathematician and inventor. "We have the largest manuscript collection of any museum in the country," he said. "It's a bit like having Archimedes's brain in a box."

Because the manuscript is delicate, scientists at nearby Johns Hopkins University will digitize the images so they can be more throughly viewed. "We're going to do an audiovisual film to tell the story of Archimedes," Mr. Vikan said, "to tell about the transition of classical knowledge to modern times."

A palimpsest is a parchment or other writing surface that has been used more than once, so that the earlier writing is only partly visible. The 174-page Archimedes palimpsest is the only manuscript containing the mathematician's "Method of Mechanical Theorems" and the original Greek version of "On Floating Bodies." Copied during the 10th century by a scribe in Constantinople, the text of Archimedes's theories was washed off in the 12th century by monks so that the parchment could be re-used. Digital technology makes it possible to read beneath the monks' writing, revealing Archimedes's text and geometrical diagrams.

Christie's catalog for the auction showed that modern technology, in the form of digital enhancement and ultraviolet photography, could clarify words and diagrams that Heiberg could not have read with the aid of only the naked eye and a magnifying glass. I expect that his translation will be improved upon, but that there will be no major surprises.

The next chapter presents the lost method, as preserved in the palimpsest, and shows how Archimedes applies it to find areas, volumes, and centers of gravity. Of course, he does not consider this technique, which involves centers of gravity, rigorous mathematics. He values it, instead, for suggesting conjectures that may turn out to be true and for enabling proofs to be found more easily.

In his letter that opens the next chapter, he offers these two benefits as reason for revealing the method to a fellow mathematician.

5

The Mechanical Method

Archimedes sent "The Method" to his friend Eratosthenes (best known today for his measurement of the circumference of the Earth and for a method of listing the prime numbers) with an introduction that reveals a good deal about his way of doing mathematics:

> Since I know that you are diligent, an excellent teacher of philosophy, and greatly interested in any mathematical investigations that may come your way, I thought it might be appropriate to write down for you a special method, by means of which you will be able to recognize certain mathematical questions with the aid of mechanics. I am convinced that this is no less useful for finding the proofs of these same theorems.
>
> Some things, which first became clear to me by the mechanical method, were afterwards proved geometrically, because their investigation by that method does not furnish an actual demonstration. It is easier to supply the proof when we have previously acquired, by the method, some knowledge of the questions than it is to find it without any previous knowledge.
>
> That is the reason why, in the case of the theorems which Eudoxus was the first to prove, namely that the volume of a cone is one-third that of the cylinder and the volume of a pyramid is one-third that of the prism with the same base and equal height, no small share of the credit should be given to Democritus, who was the first to state these facts, though without proof.
>
> I now wish to describe the method, partly because I have already spoken about it, that I may not impress some people as having uttered idle talk, and partly because I am convinced that it will prove very useful for

> mathematics. In fact, I presume there will be some among the present as well as future generations who by means of the method will be enabled to find other theorems which have not yet fallen to our share.

Volume of a Paraboloid

The simplest example of the method is finding the volume of a section of a paraboloid of revolution. The paraboloid is formed by spinning a parabola around its axis. Without any loss of generality, we may assume that the equation of the parabola is $y = x^2$ relative to suitable axes. Figure 1 shows this section, which is cut off by a plane perpendicular to the axis. It also shows the circumscribing cylinder.

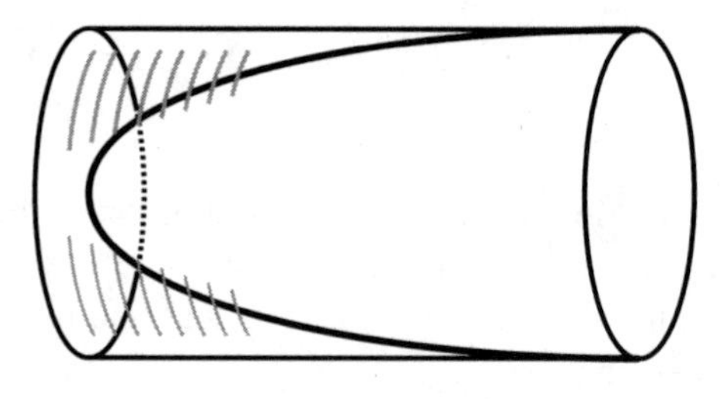

Figure 1

Archimedes illustrates his method by showing that the volume of the parabolic section is exactly half the volume of the cylinder.

Recall that one of Archimedes' three assumptions about centers of gravity is that a congruence between two objects takes one center of gravity to the other. Reflection of the cylinder into itself across the plane perpendicular to its axis and midway between its two bases is a congruence. Since it takes the center of gravity to itself, the center of gravity of the cylinder must lie on the reflecting plane. A rotation about its axis shows that the center of gravity of the cylinder also lies on its axis. Hence the center of gravity of the cylinder must be the midpoint of its axis. Similar reasoning shows that the center of gravity of the paraboloid lies somewhere on the same axis.

Let A be the vertex of the paraboloid and D the intersection of the axis with the base of the section, a disk of radius BD. The fulcrum is at A. Locate the point H on the line AD, extended, so $AD = AH$. Let K be the center of gravity of the cylinder, the midpoint of AD. The typical plane parallel to the base meets the line AD at the point S, the cylinder in a disk of radius MS, and the paraboloid in a disk of radius OS, as shown in Figure 2.

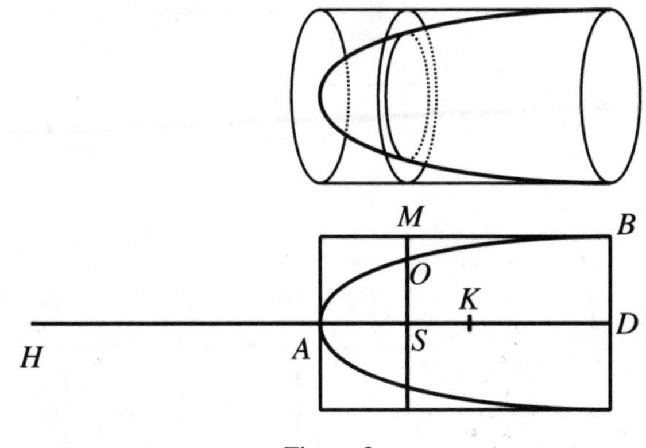

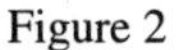

Figure 2

Since the parabola has the equation $y = x^2$ if the y-axis is HD and the x-axis the line through A perpendicular to HD,

$$\frac{BD^2}{OS^2} = \frac{AD}{AS};$$

hence,

$$\frac{MS^2}{OS^2} = \frac{AD}{AS}$$

and therefore

$$AS \cdot MS^2 = AD \cdot OS^2.$$

Thus

$$AS \cdot MS^2 = AH \cdot OS^2,$$

from which it follows that

$$AS(\pi MS^2) = AH(\pi OS^2).$$

This last equation, interpreted in physical terms, says, in view of the law of the lever, that "*the cross section of the cylinder, left where it is, balances the cross section of the paraboloid, moved so its center is at H.*" To draw this I think of moving the paraboloid so that its axis passes through H and is perpendicular to AH, as in Figure 3.

Now comes the big leap in Archimedes' mechanical method. It seems plausible that if the paraboloid in Figure 3 balances the cylinder cross section for cross section around the fulcrum A, then the two solids also balance around

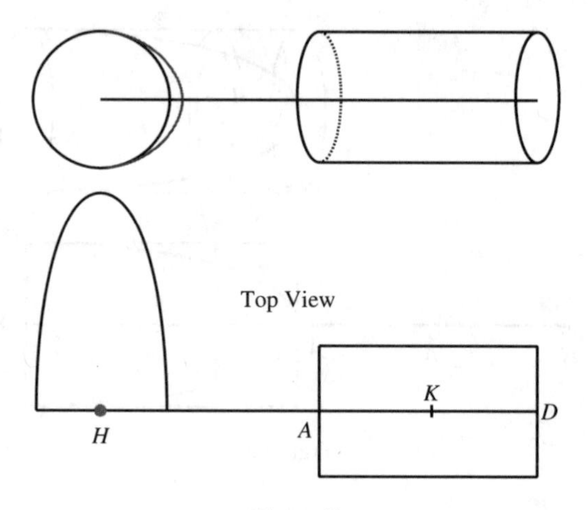

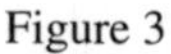
Figure 3

the A. Now the center of gravity of the paraboloid is H, and that of the cylinder is K. Since the two objects balance, and they act as though all their mass is at their centers of gravity, we have the equation

$$AH \cdot \text{volume of paraboloid} = AK \cdot \text{volume of cylinder.}$$

Consequently, since AK is half of AD and AH equals AD,

$$AD \cdot \text{volume of paraboloid} = (AD/2) \cdot \text{volume of cylinder.}$$

Thus the volume of the paraboloid is half that of the cylinder.

The big leap involves the assumption that if cross sections behave in a certain way, then the objects themselves do. A simple example shows that this is a dangerous assumption.

Consider triangle ABC in Figure 4 together with the line segment MN joining the midpoints of sides AB and AC. Let AX be a typical cross section of the triangle through A.

Each cross section AX has its center of gravity on the line MN, yet the center of gravity of triangle ABC, though made up of such lines, is not on that line. The trouble here is that the cross sections are not parallel. Calculus justifies the method when the cross sections are parallel.

Looking back at the example of the paraboloid, we see that the key is that the areas of corresponding cross sections of the two bodies are intimately related: "*The lever arm times the area of one equals a constant times the area of the other.*" That relation permits Archimedes to set up an equation involving

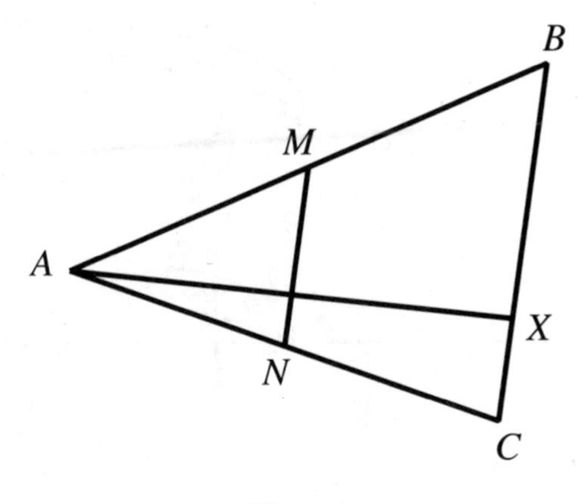

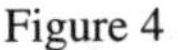

Figure 4

two volumes and two centers of gravity. So if he knows any three of those quantities he can solve for the fourth.

The method also applies to figures in the plane, where the cross sections are line segments instead of planar regions.

Center of Gravity of a Paraboloid

Archimedes observes that as a consequence of the example, the volume of the paraboloid is $\frac{3}{2}$ times the volume of the inscribed cone with the same base and the same vertex, as shown in Figure 5.

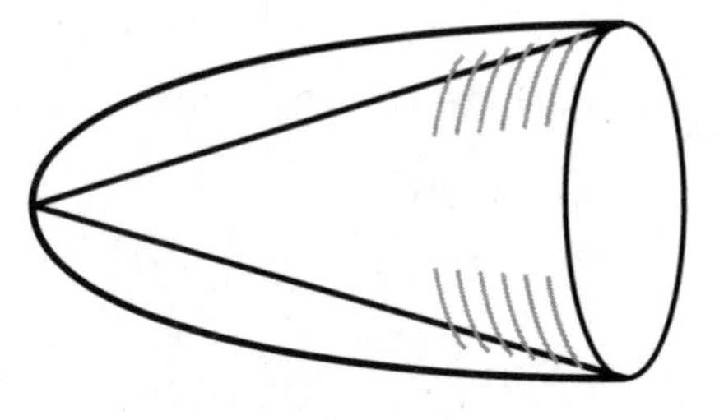

Figure 5

By comparing this cone and paraboloid, Archimedes determines the center of gravity of the paraboloid by the method. Figure 6 presents a side view of the essential elements.

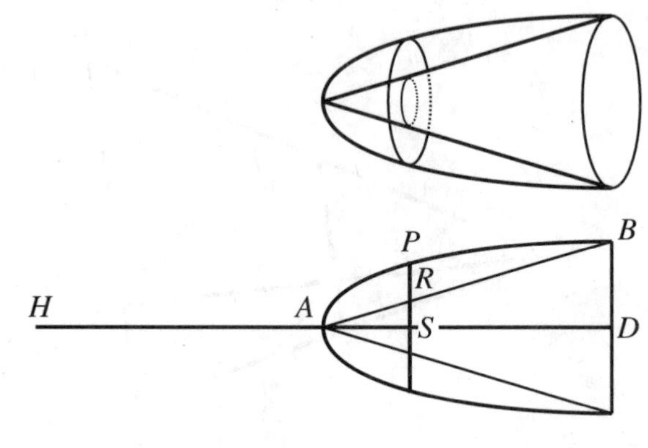

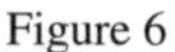
Figure 6

The common vertex is A, which serves as the fulcrum. The center of the base is D. The cross sections at the typical point S on AD have radii RS and PS. H is again placed so that $AH = AD$.

By a basic property of a parabola,

$$\frac{PS^2}{AS} = \frac{BD^2}{AD}.$$

Thus

$$AD \cdot PS^2 = AS \cdot BD^2.$$

By similar triangles,

$$\frac{RS}{AS} = \frac{BD}{AD}.$$

Eliminating BD from these last two equations gives

$$AS \cdot PS^2 = AD \cdot RS^2;$$

hence,

$$AS(\pi PS^2) = AD(\pi RS^2).$$

This last equation implies that the cross section of the paraboloid, left where it is, balances the cone's cross section placed with its center at H. Thus, according to the method, if the center of gravity of the paraboloid is at C,

$$AC \cdot \text{volume of paraboloid} = AD \cdot \text{volume of cone}.$$

Since the volume of the paraboloid is $\frac{3}{2}$ the volume of the cone, it follows that

$$AC = \frac{2}{3} \cdot AD.$$

"The center of gravity of a paraboloid of revolution is on its axis twice as far from the vertex as from the base." This information plays a key role when Archimedes studies the equilibrium of a floating object in the shape of a section of a paraboloid, as we will see in Chapter 8. In fact, in that investigation Archimedes needs the center of gravity of a section of the paraboloid cut off by a plane that is not perpendicular to its axis. He may have determined it in a lost work, called *Equilibrium*. W. Knorr, in *Archimedes' Lost Treatise on the Centers of Gravity of Solids*, listed in the references in Appendix D, discusses this work and reconstructs what may have been in it. In particular, he obtains the centers of gravity of a right circular cone and paraboloid of revolution by rigorous geometric arguments that Archimedes may have used.

Archimedes also uses the method to find the area cut off a parabola by any chord. I will give that argument, as well as his purely mathematical argument, in Chapter 7.

The Volume of a Sphere

When obtaining the volume of a sphere by his mechanical method, Archimedes uses not two solids, but three: a cone, a cylinder, and, of course, the sphere, as shown in Figure 7. The diameter of the cylinder is twice the diameter of the sphere.

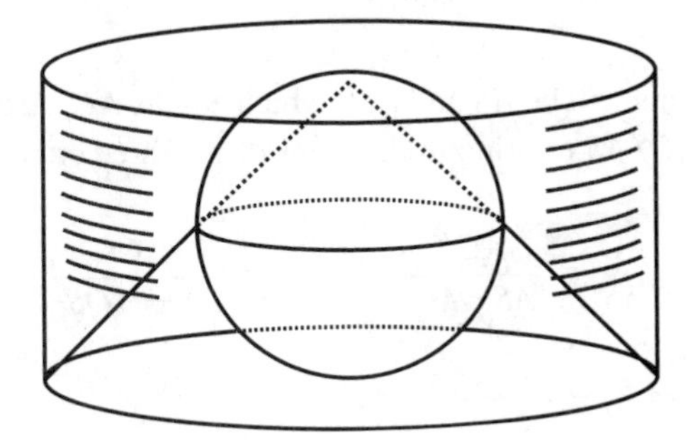

Figure 7

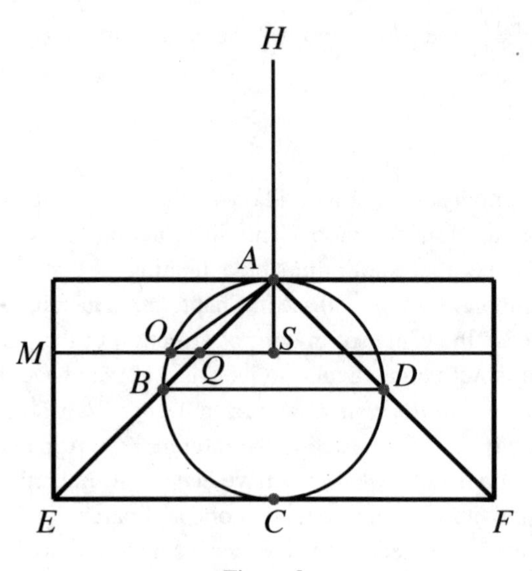

Figure 8

He considers cross sections by planes parallel to a base of the cylinder. A side view of the typical cross section is shown in Figure 8.

AC and BD are diameters of the sphere, EF a diameter of the cylinder. The typical cross section passes through M and meets the diameter AC at S. The line MS meets the sphere at O and the cone at Q. A is the midpoint of HC and will serve as the fulcrum.

Besides the Pythagorean theorem, Archimedes needs the geometric fact that $AS \cdot AC = AO^2$.

Exercise 1. Prove the fact just quoted.

The lever arms will be AS and AH, which equals AC. Archimedes manipulates the ratio AC/AS to bring in the areas of the three cross sections:

$$\frac{AC}{AS} = \frac{AC^2}{AS \cdot AC} = \frac{AC^2}{AO^2} = \frac{AC^2}{AS^2 + OS^2}.$$

Thus

$$AC(AS^2 + OS^2) = AS \cdot AC^2.$$

Since $HA = AC$, $AC = MS$, and $AS = QS$, Archimedes obtains

$$HA(QS)^2 + HA(OS)^2 = AS(MS)^2;$$

hence,

$$HA \cdot (\pi QS^2) + HA \cdot (\pi OS^2) = AS \cdot (\pi MS^2).$$

This equation implies that the cylinder, left where it is, balances the cone and sphere moved so that their centers of gravity are at H. Since the center of gravity of the cylinder is at a distance $AC/2$ from A,

$$HA \cdot \text{volume of cone} + HA \cdot \text{volume of sphere} = \frac{AC}{2} \cdot \text{volume of cylinder.}$$

Recalling that $HA = AC$ and that the cylinder is three times as large as the cone, Archimedes has

$$AC \cdot \text{volume of cone} + AC \cdot \text{volume of sphere} = \frac{AC}{2} \cdot 3 \cdot \text{volume of cone.}$$

It follows that

$$\text{volume of sphere} = \frac{1}{2} \cdot \text{volume of cone.}$$

Since the cone is eight times as large as the inscribed cone whose vertex is A and whose base is the equatorial cross section of the sphere,

$$\text{volume of sphere} = 4 \cdot \text{volume of inscribed cone.}$$

In short, as Archimedes summarizes his discovery, "The volume of the sphere is four times the volume of the cone whose height is the radius of the sphere and whose base is the equatorial cross section of the sphere."

Exercise 2. Show that the volume of a sphere is two thirds the volume of a circumscribing cylinder. (It is this result that Archimedes wanted memorialized on his gravestone.)

Archimedes then remarks, "From this theorem I conceived the notion that the surface of any sphere is four times as great as a great circle in it. For, judging from the fact that any circle is a triangle with base equal to the circumference and height equal to the radius of the circle, I suspected that in like manner, any sphere is equal to a cone with base equal to its surface and height equal to its radius."

Exercise 3. Use Archimedes' point of view and the fact that the volume of a right circular cone is one third of its height times the area of its base to show that the surface of a sphere is four times as great as its equatorial cross section.

Center of Gravity of a Hemisphere

Archimedes also finds the center of gravity of a hemisphere by his mechanical method. In his argument he assumes that the center of gravity of a cone is known, situated one quarter of the way from the base to the vertex.

He begins by introducing the same inscribed cone that played a role in the previous argument. Figure 9 shows the situation, including the typical cross-sectional plane.

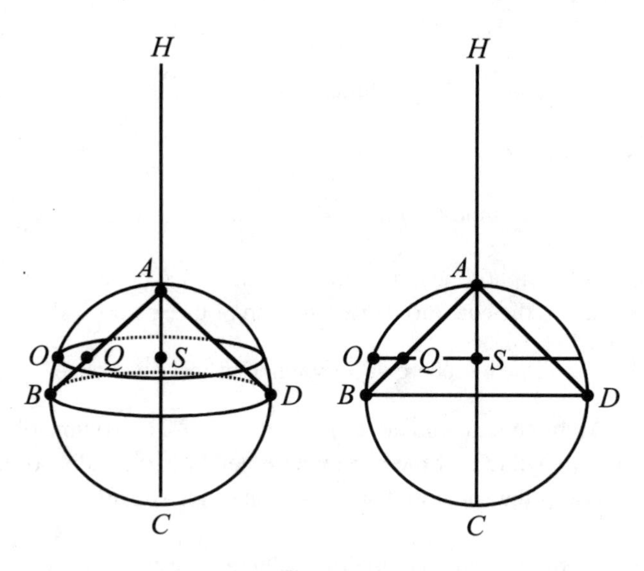

Figure 9

He determines the center of gravity of the top hemisphere.

The center of gravity of the top hemisphere, G, lies somewhere on the radius through A.

Again he starts with the ratio AC/AS, but this time multiplies by AS:

$$\frac{AC}{AS} = \frac{AC \cdot AS}{AS^2} = \frac{AO^2}{AS^2} = \frac{AS^2 + OS^2}{AS^2} = \frac{QS^2 + OS^2}{QS^2}.$$

Thus, since $AC = HA$,

$$HA \cdot QS^2 = AS \cdot QS^2 + AS \cdot OS^2;$$

hence,

$$HA \cdot (\pi QS^2) = AS \cdot (\pi QS^2) + AS \cdot (\pi OS^2).$$

This equation implies, according to the method, that the cone placed with its center of gravity at H balances the cone and the hemisphere left where they are. Now, the center of gravity of the cone is at a distance $\frac{3}{4}$ of the radius from A, hence at a distance $\frac{3}{8}AC$ from A. Since $HA = AC$, Archimedes has the equation

$$AC \cdot \text{volume of cone} = \frac{3AC}{8} \cdot \text{volume of cone} + AG \cdot \text{volume of hemisphere}.$$

Hence,

$$\frac{5}{8} \cdot AC \cdot \text{volume of cone} = AG \cdot 2 \text{ volume of cone},$$

from which it follows that

$$\frac{AG}{AC} = \frac{5}{16}.$$

Since AC is twice the radius, AG is $\frac{5}{8}$ of the radius. The center of gravity of a hemisphere divides a radius into two segments such that the length of the segment adjacent to the surface is to the length of the segment adjacent to the base as five is to three.

Archimedes does not consider any of these arguments rigorous, convincing though they may be. He still feels obliged to find the truly geometric arguments that do not appeal to centers of gravity and balancing weights. For example, in Chapter 10 we will see his rigorous proofs that obtain the surface area and volume of a sphere.

6

Two Sums

In the next chapter Archimedes finds the area of a segment of a parabola cut off by any of its chords. In his proof he needs the formula for the sum of a geometric series. In Chapter 9, where he finds the area within a spiral, he needs a formula for the sum of the first n squares. In order not to disrupt the flow of thought in those chapters, we obtain the two formulas now.

Sum of a Geometric Series

Archimedes needs to determine how the following sum behaves as n increases:

$$1 + \frac{1}{4} + \frac{1}{4^2} + \cdots + \frac{1}{4^{n-1}} + \frac{1}{4^n}. \tag{1}$$

His key is the equation

$$1 + \frac{1}{4} + \frac{1}{4^2} + \cdots + \frac{1}{4^{n-1}} + \frac{1}{4^n} + \frac{1}{3} \cdot \frac{1}{4^n} = \frac{4}{3}. \tag{2}$$

Since $1/(3 \cdot 4^n)$ approaches 0 as n increases, this equation implies that the sum (1) approaches $\frac{4}{3}$ as n increases.

To show that (2) holds, note that

$$\frac{1}{4^n} + \frac{1}{3} \cdot \frac{1}{4^n} = \frac{1}{4^n}\left(1 + \frac{1}{3}\right) = \frac{1}{4^n} \cdot \frac{4}{3} = \frac{1}{3 \cdot 4^{n-1}}.$$

Thus, adding $(1/3)(1/4^n)$ to $1/4^n$ replaces the last two summands in (2) by one term, $(1/3)(1/4^{n-1})$.

Applying this operation once changes the left side of (2) to

$$1 + \frac{1}{4} + \frac{1}{4^2} + \cdots + \frac{1}{4^{n-1}} + \frac{1}{3 \cdot 4^{n-1}}.$$

The operation simply lowers the largest exponent, n, to $n - 1$. Applying the operation repeatedly reduces the left side of (2) finally to

$$1 + \frac{1}{3 \cdot 4^0} = \frac{4}{3}.$$

This establishes (2).

Exercise 1. Let c be greater than 1. Use the same approach to show that $1 + 1/c + 1/c^2 + \cdots + 1/c^n$ approaches $c/(c - 1)$ as n increases.

Over the centuries mathematicians have found many ways to show that the sum (1) approaches $\frac{4}{3}$ as n increases. Three of them are described in my book, *Strength in Numbers*. The February 1999 issue of *Mathematics Magazine* offered yet another approach, a "Proof without Words" due to Rick Mabry. It involves gazing awhile at Figure 1, which shows an endless string of equilateral triangles inside an equilateral triangle whose area is 4.

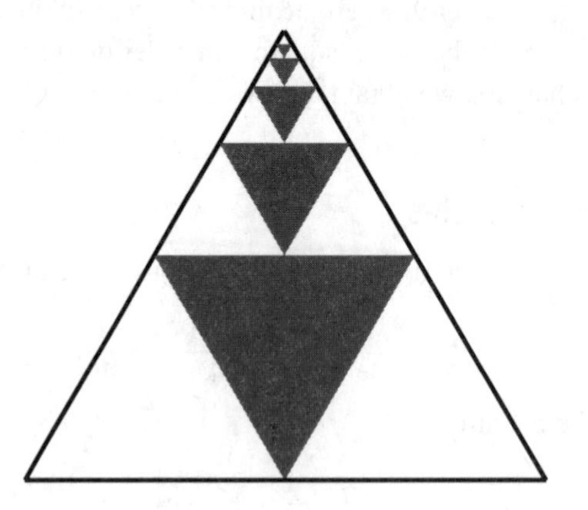

Figure 1

Sum of Squares

Archimedes also needs a short formula for the sum of the first n squares,

$$S_n = 1^2 + 2^2 + 3^2 + \cdots + n^2. \tag{3}$$

He first finds the sum of the first n whole numbers by the method every algebra student meets today:

Let

$$T_n = 1 + 2 + 3 + \cdots + (n-1) + n.$$

Then he also has

$$T_n = n + (n-1) + \cdots + 3 + 2 + 1.$$

Adding the two versions of the sum, column by column, yields n sums, each equal to $n+1$; hence,

$$2T_n = (n+1)n,$$

and therefore

$$T_n = \frac{n(n+1)}{2}.$$

The preceding proof is based on the equations $n = 1 + (n-1)$, $n = 2 + (n-2), \ldots, n = (n-1) + 1$. Archimedes' argument for the sum of squares resembles the proof just given. He begins with n equations:

$$\begin{aligned}
n^2 &= (1 + (n-1))^2 = 1^2 + (n-1)^2 + 2 \cdot 1 \cdot (n-1) \\
n^2 &= (2 + (n-2))^2 = 2^2 + (n-2)^2 + 2 \cdot 2 \cdot (n-2) \\
&\cdots\cdots\cdots\cdots\cdots\cdots\cdots\cdots\cdots \\
n^2 &= ((n-1) + 1)^2 = (n-1)^2 + 1^2 + 2 \cdot (n-1) \cdot 1 \\
2n^2 &= n^2 + n^2.
\end{aligned}$$

He has to toss in this last equation both because his numbers represent lengths of line segments and because he does not have a zero at his disposal. Adding the n rows, column by column, gives

$$(n+1)n^2 = S_n + S_n + 2(n-1) + 4(n-2) + 6(n-3) + \cdots + (2n-2)1.$$

Let

$$Q = 2(n-1) + 4(n-2) + 6(n-3) + \cdots + (2n-2)1.$$

It would appear that Archimedes now faces a more ungainly sum than the one he started with. Nevertheless he manages to obtain a short formula for Q and therefore for S_n.

At this point he pulls a rabbit out of the hat, adding T_n to Q,

$$T_n + Q = 1 + 2 + \cdots + n + 2(n-1) + 4(n-2) + \cdots + (2n-2)1,$$

which he rewrites as

$$T_n + Q = n + 3(n-1) + 5(n-2) + \cdots + (2n-1)1.$$

Using another ingenious device, he will show that this forbidding sum is simply S_n in disguise. This fact would not surprise us if we drew the pyramidal diagram, Figure 2, that represents, S_4.

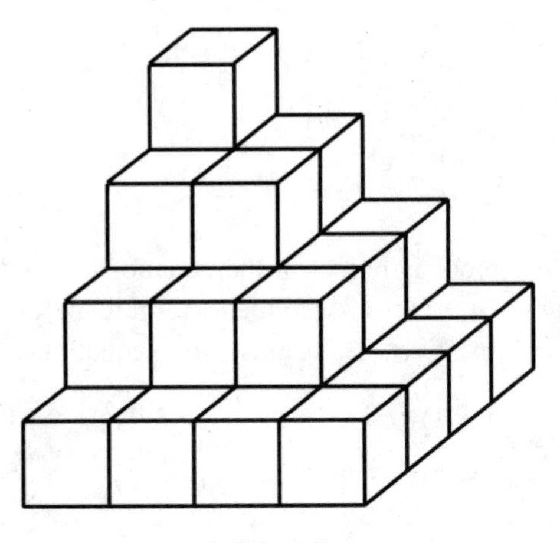

Figure 2

Adding up the number of little cubes by vertical columns rather than by horizontal slabs shows that

$$1^2 + 2^2 + 3^2 + 4^2 = 4 + 3(3) + 5(2) + 7(1).$$

Archimedes does not appeal to such a diagram. Instead he goes on as follows:

$$\begin{aligned}
n^2 &= n + (n-1)n = n + 2(1 + 2 + \cdots + (n-1)) \\
(n-1)^2 &= n - 1 + (n-2)(n-1) = n - 1 + 2(1 + 2 + \cdots + (n-2)) \\
(n-2)^2 &= n - 2 + (n-3)(n-2) = n - 2 + 2(1 + 2 + \cdots + (n-3)) \\
&\cdots\cdots\cdots\cdots\cdots\cdots\cdots\cdots \\
2^2 &= 2 + 1 \cdot 2 = 2 + 2(1) \\
1^2 &= 1 = 1.
\end{aligned}$$

Adding all the terms while collecting like summands, he obtains

$$S_n = n + 3(n-1) + 5(n-2) + \cdots + (2n-1)1.$$

Thus

$$T_n + Q = S_n.$$

Combining this equation with his knowledge that $(n+1)n^2 = 2S_n + Q$, Archimedes deduces that

$$T_n + (n+1)n^2 = 2S_n + S_n.$$

Thus

$$\frac{n(n+1)}{2} + (n+1)n^2 = 3S_n.$$

A little algebra changes this into a formula for the sum of squares:

$$S_n = \frac{n^3}{3} + \frac{n^2}{2} + \frac{n}{6}.$$

Exercise 2. Supply the little algebra.

Actually, all that Archimedes needs to know is that S_n is greater than $n^3/3$ and that S_{n-1} is less than $n^3/3$. The first is clear, as a glance at the formula for S_n shows. To establish the second inequality, subtract n^2 from the formula for S_n, obtaining

$$S_{n-1} = \frac{n^3}{3} - \frac{n^2}{2} + \frac{n}{6}.$$

Since $n^2/2$ is larger than $n/6$, it follows that S_{n-1} is less than $n^3/3$.

There is no evidence that Archimedes ever tried to estimate the sum of the first n cubes. I tried to use his method to get a short formula for this sum, but the expressions I met became too messy. The Arabic mathematician Alhazan, around the year 1000 A.D, did develop formulas for the sum of cubes and the sum of fourth powers in order to find certain volumes. His method is quite different, for it involves cutting up a rectangle into smaller rectangles, whose dimensions are determined by powers and sums of powers. This method is described in *The Historical Development of Calculus*, by C. H. Edwards, Jr., cited in Appendix D. Still I wonder how Archimedes would have treated the sum of cubes if he had ever needed it.

7

The Parabola

In his introduction to *Quadrature of the Parabola*, Archimedes writes to Dositheus,

> When I heard that Conon, who was my friend, was dead, but that you were acquainted with Conon and versed in geometry, while I grieved for the loss not only of a friend but of an admirable mathematician, I set myself the task of communicating to you, as I had intended to send to Conon, a certain theorem which had not been investigated before but has now been investigated by me. I discovered it first by means of mechanics and then exhibited it by means of geometry.
>
> Some of the earlier geometers tried to prove it possible to find a rectilinear area equal to a given circle. After that they tried to square the area bounded by a conic section and a line, assuming lemmas not easily conceded, so that it was recognized by most people that the problem was not solved. But I am not aware that any one of my predecessors has tried to square the section bounded by a parabola and a line. I have now discovered the solution of this problem.
>
> I have written out the proof and send it to you, first as investigated by means of mechanics, and afterwards as demonstrated by geometry. Prefixed are also the elementary propositions about conics which are of service in the proof. Farewell.

This is what Archimedes discovered:
Consider the region bounded by a parabola and a chord AC, as in Figure 1.

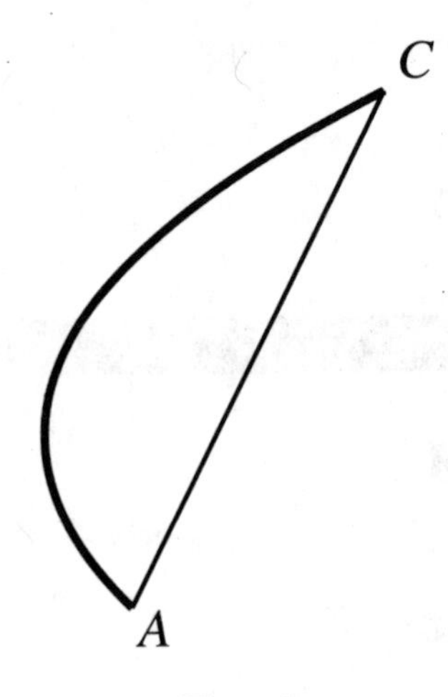

Figure 1

Let B be the point on the parabola where the tangent is parallel to the chord AC, as in Figure 2.

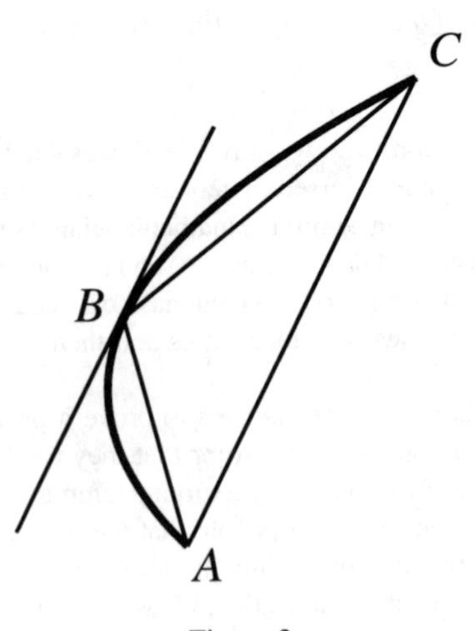

Figure 2

Archimedes shows that the area of the parabolic section is $\frac{4}{3}$ times the area of triangle ABC, whose base is the chord and whose third vertex is B. In this way he manages to express the parabolic area in terms of a "rectilinear area." In Archimedes' time, the phrase "find the area" meant "express the area in terms of the area of a simpler region, preferably a polygon."

In this chapter we will see his proof by mechanics and then his rigorous geometric proof. After that we will watch as he finds the center of gravity of a parabolic section.

The Area by Mechanics

Archimedes begins by constructing a much larger triangle than the one in Figure 2. In Figure 3, FC is the tangent at C to the parabola and D is the midpoint of AC. ED is the axis of the section. (See Appendix A for the definition of the axis of a section.) AF is parallel to the axis. MO is a typical cross section of triangle AFC parallel to the axis. MO crosses CH at N and the parabola at P. H is located on the extension of CK so that HK equals KC. (K will be the fulcrum in his argument.)

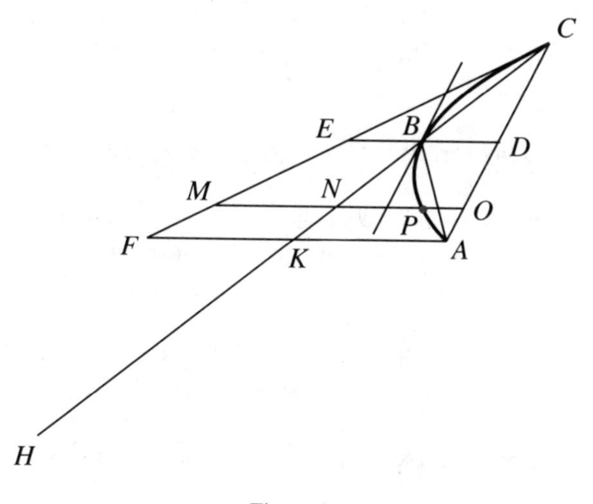

Figure 3

Archimedes uses only two facts about a parabola: that $MO/OP = CA/AO$ and that $BD = DE/2$. Both are established in Appendix A. He uses only the first one to prove, as his first step, that the area of the parabolic section is $\frac{1}{3}$ times the area of the large triangle AFC.

His argument is brief. He starts with the equation

$$\frac{MO}{OP} = \frac{CA}{AO}.$$

Then, since ON is parallel to AK,

$$\frac{CA}{AO} = \frac{CK}{KN}.$$

But by the definition of H, $CK = HK$. Therefore, Archimedes has

$$\frac{MO}{OP} = \frac{HK}{KN};$$

hence,

$$MO \cdot KN = OP \cdot HK.$$

This last equation implies that "the cross section MO of the big triangle, where it is, balances the cross section of the parabolic section placed at H." By the mechanical method, Archimedes concludes that triangle CAF, left where it is, balances the parabolic section at H. The center of gravity of the triangle is on the median CK, one-third of the way from K to C, as mentioned in Chapter 3. Hence its lever is $CK/3$. The lever for the parabolic section, placed at H, is $HK = CK$. Thus, he has

$$\left(\frac{CK}{3}\right) \cdot \text{area of triangle } CAF = CK \cdot \text{area of parabolic section.}$$

Consequently the area of the parabolic section is a third the area of triangle CAF. All that remains is to show that the area of triangle CAF is four times the area of triangle ABC. That is where the equation $BD = DE/2$ comes in.

Exercise 1. Prove that triangle CAF is four times as large as triangle ABC. (Hint: A median of a triangle bisects the area.)

All told, the parabolic section is $\frac{4}{3}$ as large as the inscribed triangle ABC.

The Area by Pure Geometry

Archimedes' completely geometric proof stays inside the parabolic section, whose area he approximates by inscribed polygons made of triangles. His first approximation is the triangle ABC. The "error" at this point is the area of the two shaded parabolic sections shown in Figure 4.

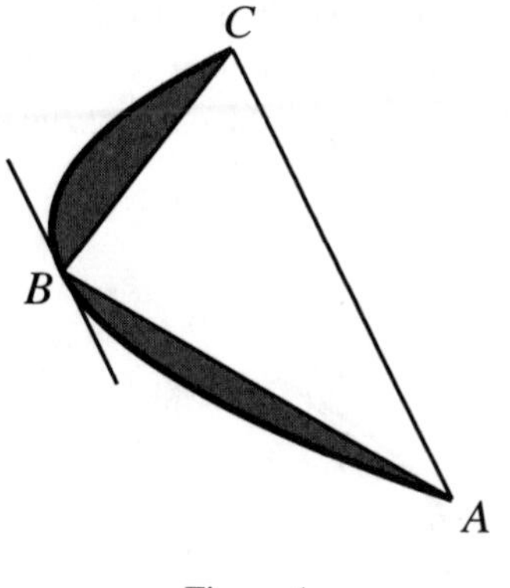

Figure 4

At his first stage he inscribes two smaller triangles, one in each of these two sections, defined in the same way that he defined triangle ABC. These two triangles are shaded in Figure 5.

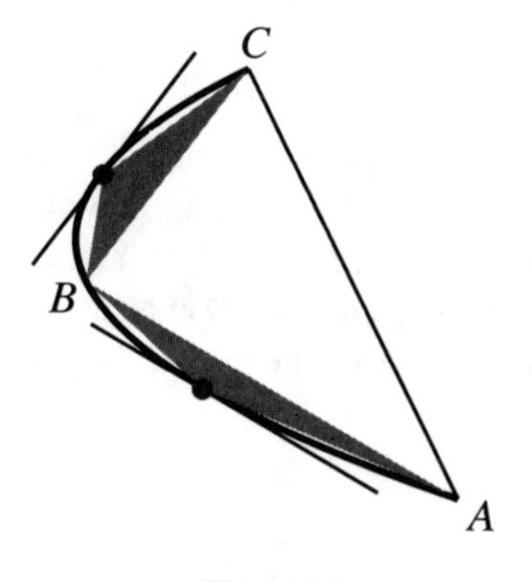

Figure 5

At the next stage he repeats the process, adding four even smaller triangles. He continues this process, adding 2^n triangles at the nth stage. He then shows that the total area of the triangles added at each stage is a quarter of the total area of the ones added in the preceding stage. Therefore the total area of all the triangles added in the various stages is represented by a geometric series. Finally, he shows that the "error" approaches zero as the number of stages increases.

Now for the details.

Consider the section of a parabola cut off by the chord AC. Let B be the point on the parabola where the tangent is parallel to AC and let D be the midpoint of AC. Let P be the point on the parabola where the tangent is parallel to BC, as shown in Figure 6.

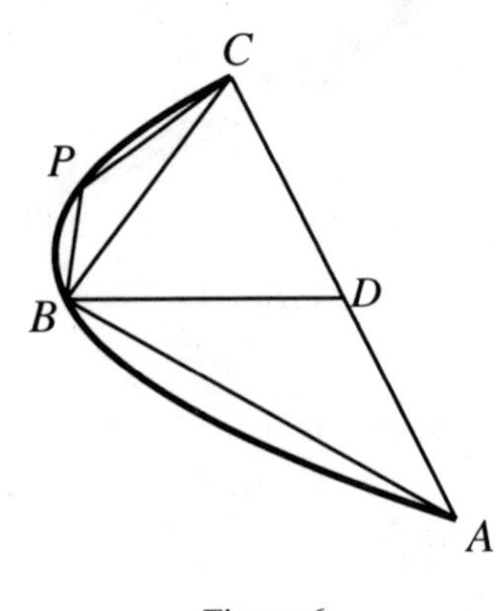

Figure 6

The key to Archimedes' argument is that the area of triangle PBC is one-quarter the area of triangle BDC. To begin, he draws the line through P parallel to the axis BD. It meets DC at M and BC at Y. He also constructs the line through P parallel to CD, which meets BD at N, as in Figure 7. Note that M is the midpoint of CD, since PY is on the axis of the parabolic section BPC.

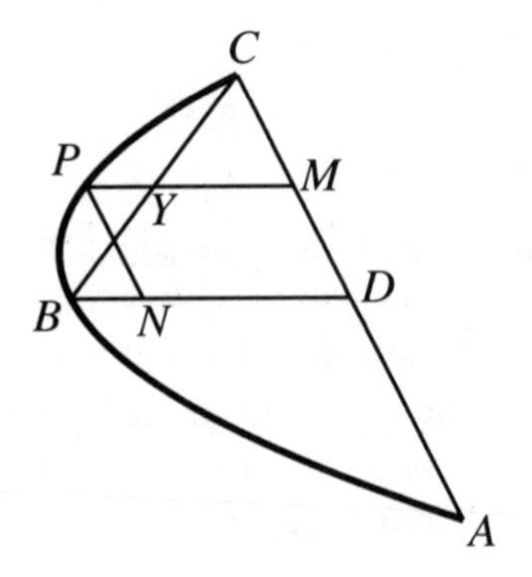

Figure 7

Archimedes shows first that $YM = 2PY$, as follows. Since $CD = 2PN$ and

$$\frac{BD}{BN} = \frac{CD^2}{PN^2},$$

he has $BD/BN = 4$, hence $BD = 4BN$. Thus $PM = 3BN$. Also, $YM = \left(\frac{1}{2}\right) BD = 2BN$. Consequently, $PY = 3BN - 2BN = BN$. Hence, $YM = 2PY$, as he claims.

With the information that $YM = 2PY$, Archimedes easily shows that the area of triangle BPC is one-quarter of the area of triangle BDC. To do this he introduces the line BM, as in Figure 8.

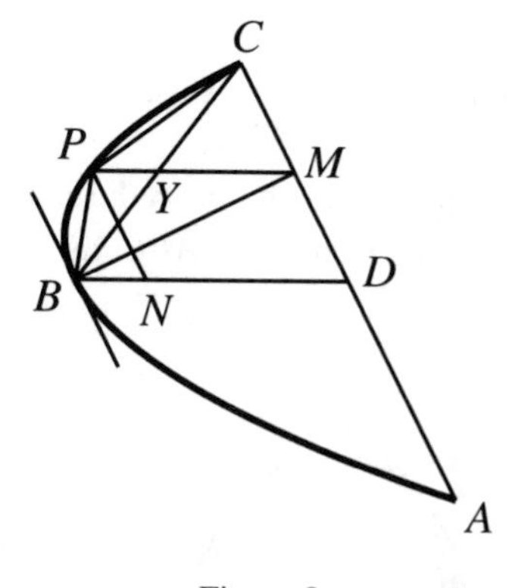

Figure 8

Since M is the midpoint of CD, the area of triangle BDC is twice the area of triangle BMC. Then, since $YM = 2PY$, the area of triangle BMC is twice the area of triangle BPC. All told, then, the area of triangle BCD is four times the area of triangle BPC.

In short,

$$\text{area of triangle } BPC = \left(\frac{1}{4}\right) \text{area of triangle } BCD.$$

Exercise 2. Justify the assertion that the area of triangle BMC is twice the area of triangle BPC.

Now Archimedes begins to fill out the parabolic section with triangles. At the first stage he adds the two shaded triangles in Figure 9. The tangent at P is parallel to BC and the tangent at Q is parallel to AB.

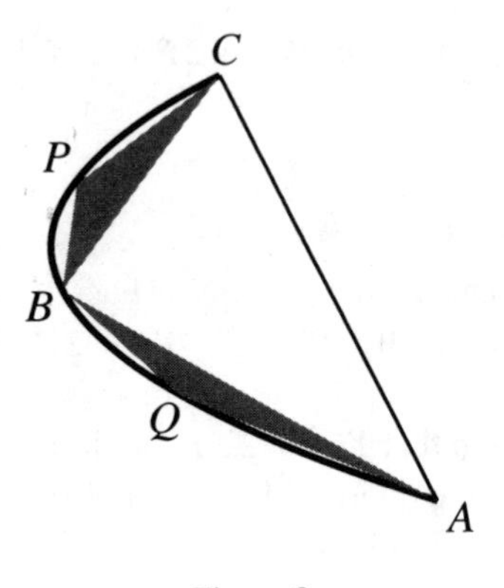

Figure 9

The total shaded area in the two little triangles is one-quarter the area of triangle ABC.

At the second stage, he repeats the procedure, with each of the triangles AQB and BPC playing the role of triangle ABC, as shown in Figure 10.

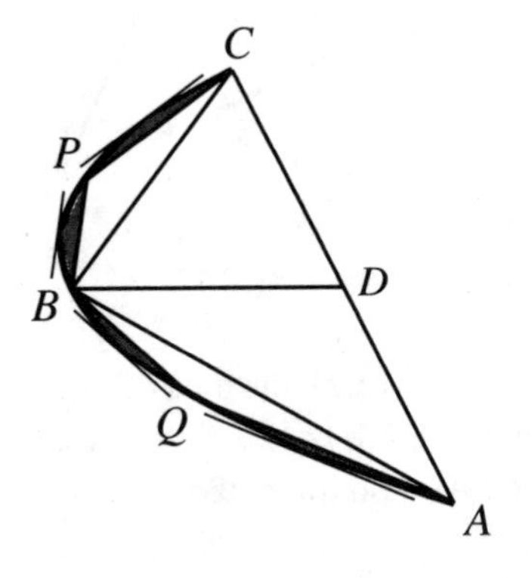

Figure 10

The total area of the four triangles added at this second stage is one-quarter of the total area of the two triangles added at the first stage. At the third stage he adds eight triangles in a similar manner; their total area is one-quarter of the area added at the second stage. After n stages, the total area of the polygon formed from all the triangles constructed is

$$\left(1 + \frac{1}{4} + \frac{1}{4^2} + \cdots + \frac{1}{4^n}\right) \cdot \text{area of triangle } ABC.$$

As Archimedes shows (see Chapter 6), the sum of the geometric series approaches $\frac{4}{3}$ as n increases. Thus the area of the parabolic section is at least $\frac{4}{3}$ times the area of triangle ABC.

All that remains is to show that the error in the polygonal approximation approaches zero as n increases. This error is the area of the part of the parabolic section outside the polygonal approximation made up of all the triangles constructed up through the nth stage.

The region not covered at any stage is made up of little parabolic sections. Figure 11 shows a typical piece cut off by a chord FG.

Figure 11

At the next stage a triangle FGH is added, with H the point on the parabola where the tangent is parallel to FG, as in Figure 12. The error is reduced to the shaded region shown in this figure.

Construct the parallelogram whose sides are FG, the tangent at H, and the lines through F and G parallel to the axis of the section, as shown in Figure 13.

Since the area of triangle FGH is half the area of the parallelogram, it is more than half the area of the parabolic section. Consequently, the error at each stage is less than half the error at the previous stage. It follows that the error does indeed approach zero as the number of stages increases.

Exercise 3. Knowing that the area of a parabolic section is $\frac{4}{3}$ times the area of the inscribed triangle ABC of Figure 4, find the ratio between the error at the $(n + 1)$st stage and the error at the nth stage. (It was just shown that this ratio is less than $\frac{1}{2}$.)

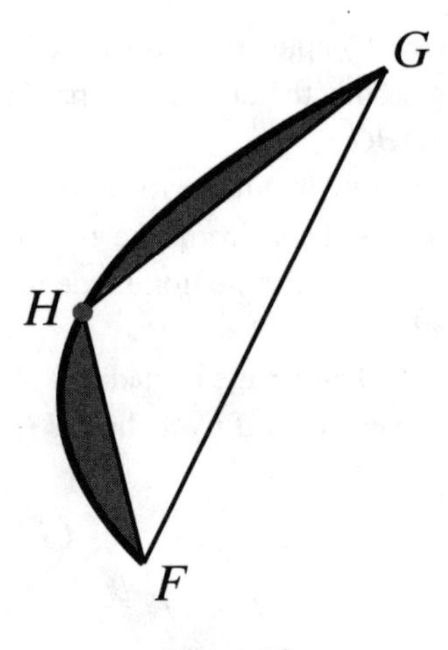

Figure 12

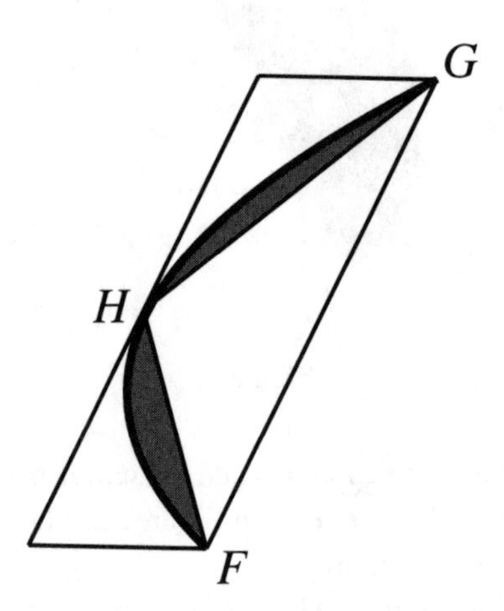

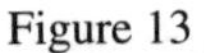
Figure 13

This is Archimedes' rigorous geometric method for finding the area of a parabolic section. His proof meets the standards of his day, when the concept of "area" was taken as given. Not until the end of the nineteenth century did mathematicians bother to define area and consider such questions as, "Does every bounded set in the plane have an area? If not, which ones do and which don't?" Peano and Jordan were the first mathematicians to provide a mathematical foundation for the notion of area.

The Center of Gravity of a Parabolic Section

In his work *Equilibrium of Planes*, Archimedes determines the center of gravity of a parabolic section. In a series of propositions whose proofs use the same polygons constructed in the preceding proof, he proves that the center of gravity lies on the axis of the section and its relative position on the axis is the same for all sections. Figure 14 shows this information.

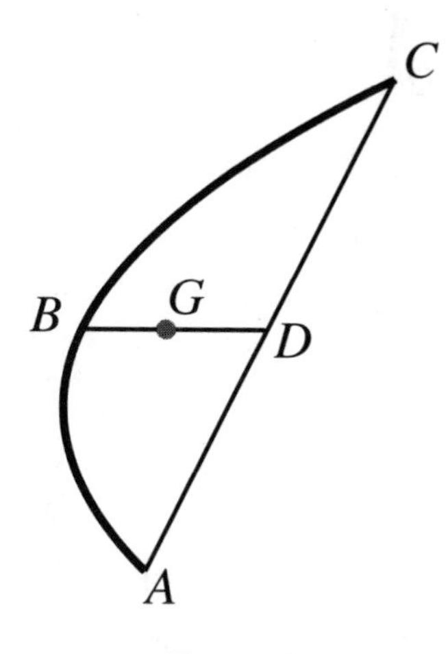

Figure 14

BD is the axis of the section, which means that the tangent at B is parallel to AC and D is the midpoint of AC. G is the center of gravity of the section.

Let $GD = r\ BD$. Archimedes determines the unknown ratio r indirectly, by setting up an equation for it. To do this, he draws triangle ABC, with center of gravity F. The two sections left over, BPC and AQB, have respective centers of gravity G' and G'', as shown in Figure 15.

The big triangle ABC balances the two sectors around the fulcrum G. After finding the lever arms and areas of these three regions, Archimedes obtains an equation for r.

Let the area of the parabolic section ABC be, say, s. Then the area of triangle ABC is $\left(\frac{3}{4}\right) s$ and the total area of the two little sectors left over is $\left(\frac{1}{4}\right) s$. Since the centers of gravity of the section ABC and the triangle ABC both lie on BD, so does the center of gravity of the union of the two little sections.

Since F is one-third of the way from D to B, $DF = \left(\frac{1}{3}\right) BD$. Thus the lever arm of the large triangle is $DG - DF = r\ BD - \left(\frac{1}{3}\right) BD$. The lever arm

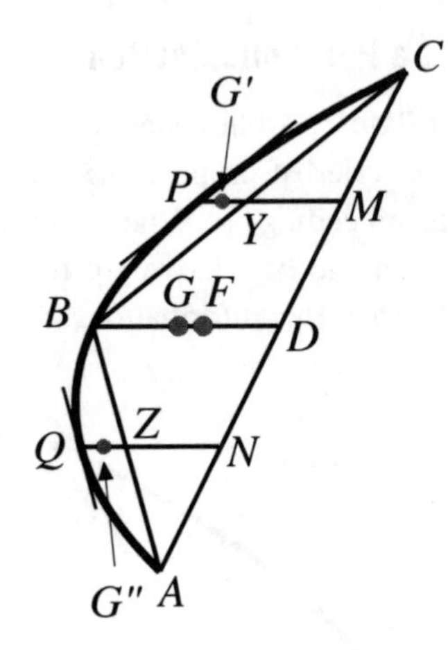

Figure 15

of the two small sectors is $MG' - DG$. Now,

$$MG' = MY + YG' = \frac{1}{2} \cdot BD + rPY = \frac{1}{2} \cdot BD + \frac{r}{4} \cdot BD,$$

since, as was shown earlier in this chapter, $PY = \left(\frac{1}{4}\right) BD$. Hence the lever arm of the two small sections is

$$\frac{1}{2} \cdot BD + \frac{r}{4} \cdot BD - r\, BD.$$

Since G is the fulcrum around which the big triangle ABC balances the two small sections, Archimedes has the equation

$$\underbrace{\left(\frac{1}{2} + \frac{r}{4} - r\right) BD}_{\text{arm}} \underbrace{\left(\frac{s}{4}\right)}_{\text{area}} = \underbrace{\left(r - \frac{1}{3}\right) BD}_{\text{arm}} \underbrace{\left(\frac{3s}{4}\right)}_{\text{area}}.$$

This implies that

$$\frac{1}{2} - \frac{3r}{4} = 3r - 1.$$

It follows that $r = \frac{2}{5}$. The center of gravity of a parabolic section lies on its axis two-fifths of the way from its base to its vertex. This fact plays an important role in the study of the equilibrium of a ship whose hull has a parabolic cross section, as the next chapter will show.

8

Floating Bodies

Imagine sawing a wood ball into two pieces. Place one of the pieces in water, with its flat base parallel to the water's surface, as in Figure 1.

Figure 1

Then tilt it but keep its base above the water, as in Figure 2.

Figure 2

When you let go will it tip over or will it return to the vertical position of Figure 1? Does the answer depend on how much is cut off?

For Archimedes this is just a warm-up exercise, which we will do in a moment. What really interests him is the following much more challenging problem.

Imagine, instead, that you have a section of a wood paraboloid of revolution floating with its base parallel to the water's surface, as in Figure 3.

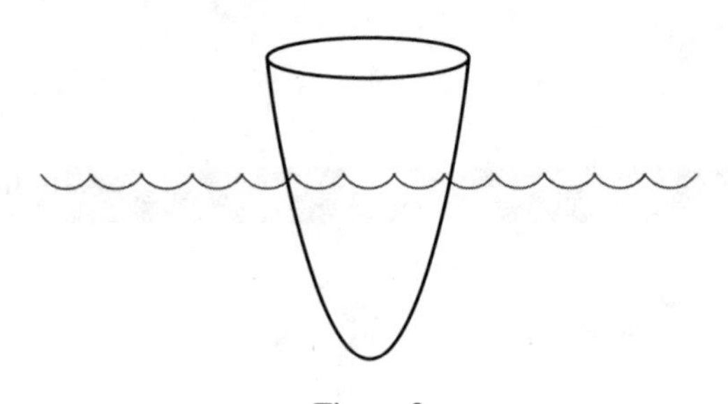

Figure 3

If you tilt it, but keep the base above the water, will it return to the vertical or will it fall over? The section may be tall, as in Figure 3, or shallow, like a saucer, as in Figure 4.

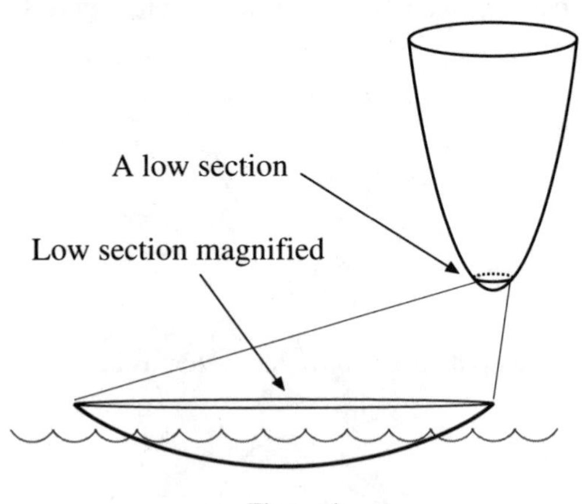

Figure 4

Archimedes investigates questions like these in his *On the Equilibrium of Floating Bodies*. His answers depend, in the case of the paraboloid, on the section's height and the wood's density.

I know of no practical reason why Archimedes raised these questions, for neither the ball nor the paraboloid resembles a ship's hull. A parabolic prism would be a fair approximation of a loaded ship, but there is no surviving evidence that he studied this case (though he had all the mathematics he would have needed). I think he was motivated by curiosity, which has been the main driving force in the expansion of our knowledge of the universe.

Dijksterhuis says of this research, "We have now come to that part of Archimedes' work which deserves the highest admiration of the present-day mathematician, both for the high standard of the results obtained, which would seem to be quite beyond the pale of classical mathematics, and for the ingenuity of the arguments."

I agree, for in this work, where Archimedes single-handedly initiates the discipline of naval architecture, we see a masterly application of pure mathematics and physical intuition. It is the fruit of much of his geometry and his work on centers of gravity.

I will first collect the geometric facts and physical principles on which he bases his analysis.

Background

Consider a line perpendicular to a parabola (a "normal") at a point P, not the vertex, on the parabola. It meets the axis of the parabola at a point C. The line through P that is perpendicular to the axis meets the axis at a point O, as in Figure 5.

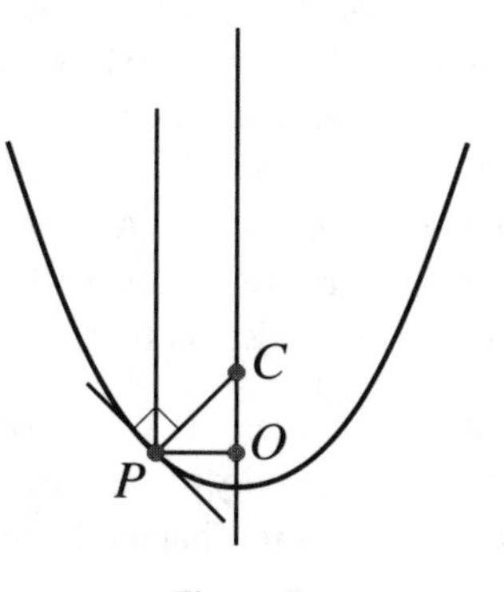

Figure 5

As was well-known in Archimedes' time, the length of CO is independent of the choice of the point P and equals twice the distance from the focus to the vertex (see Appendix A). If f is the distance from the focus to the vertex, OC is $2f$, a quantity that plays a critical role in this analysis.

Next, consider a section of a paraboloid of revolution cut off by a plane not necessarily perpendicular to its axis. Let P be the point of contact of the tangent plane parallel to the base. The line through P, parallel to the axis of the paraboloid, meets the base at a point V. (V is the center of the elliptical base, but this fact will not be needed.) PV is called the "axis" of the section. The vertex of the paraboloid is A. These points are shown in Figure 6 in perspective and side view.

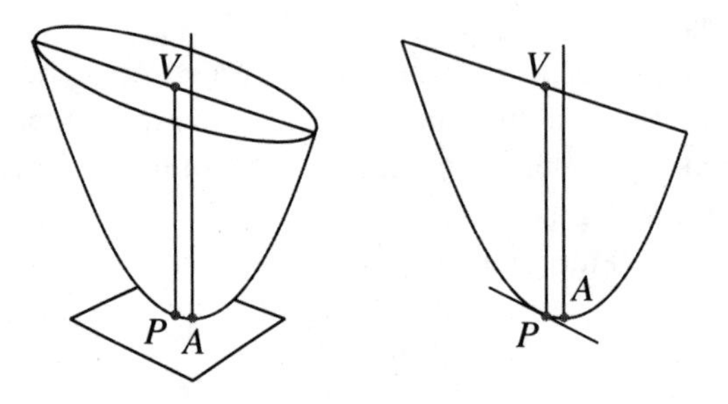

Figure 6

Archimedes proves, in a long series of propositions in his book *On Conoids and Spheres*, that the volume of such a section is proportional to PV^2. A modern proof of this is sketched in Appendix A.

Archimedes also uses the fact that the center of gravity of a section lies on its axis, PV, and is two-thirds of the way from P to V. As we saw in Chapter 5, he establishes this by the mechanical method when the base is perpendicular to the axis of the paraboloid. However, he remarks in his work *On Floating Bodies* that "it was proved in *On Equilibria*." As mentioned in Chapter 5, Knorr offers a rigorous proof that may have appeared in that lost work.

Archimedes bases his analysis also on four physical principles:

1. If an object is cut into two pieces, its center of gravity lies on the line segment joining the centers of gravity of the pieces. (The precise location is described by the law of the lever in Chapter 2, but it will not be needed.)
2. Any solid less dense than water will, if placed in water, be so far immersed that the weight of the solid will equal the weight of the displaced water.

   The denser the solid, the more of it will be submerged. To be specific, say that the volume of the solid is S and the volume of displaced water is W. Assume that the density of water is 1 (1 gram per cubic centimeter) and the density of the solid, which we may think of as made of wood, is d. Then the weight of the solid is dS and that of the displaced water is $1W = W$. Hence,

$$dS = W$$

or

$$d = W/S.$$

This equation says that "The density of the solid is equal to the volume of water displaced divided by the volume of the solid." Note that for a floating body, d is less than 1 and that the closer d is to 1, the deeper the solid sinks in the water.

3. A solid immersed in water is driven up by a force equal to the difference between its weight and the weight of the displaced water.

   Archimedes develops these three principles from simpler assumptions. However, the next one he treats as an axiom.

4. A body forced upward in water is forced upward along the perpendicular to the surface of the water that passes through the center of gravity of the submerged part.

The Technique

Imagine a body partially submerged in water. Let C be its center of gravity. The center of gravity of the submerged part is F, and H is the center of gravity of the "dry" part, exposed above the water. These three points are shown in Figure 7.

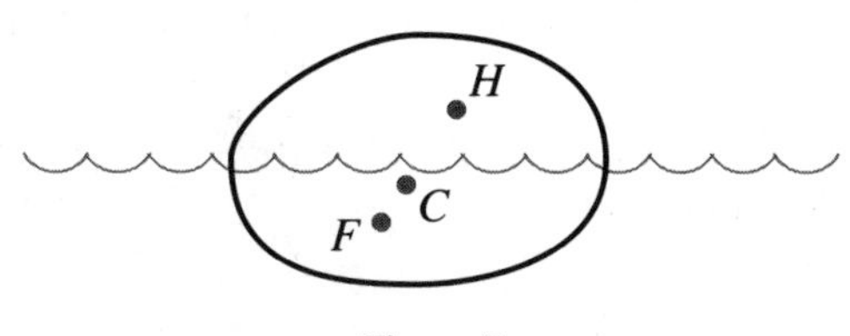

Figure 7

Archimedes pictures gravity acting on the exposed part as if all of its mass were at H. Similarly, gravity acts on the submerged part as if that part were all at F. The water pushes directly up as though it were all pressing against F. Since this force equals the weight of the entire solid, it is greater than the downward force of gravity at F on the submerged part. These three forces are shown in Figure 8.

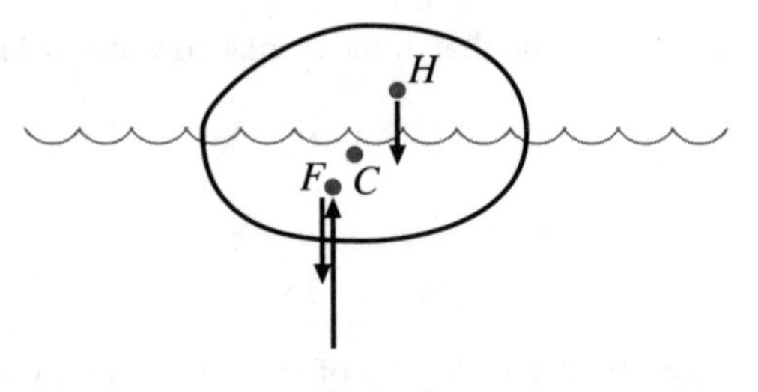

Figure 8

The net effect at F is an upward force. We may therefore reduce the situation to two forces, an upward one at F and a downward one at H, as in Figure 9.

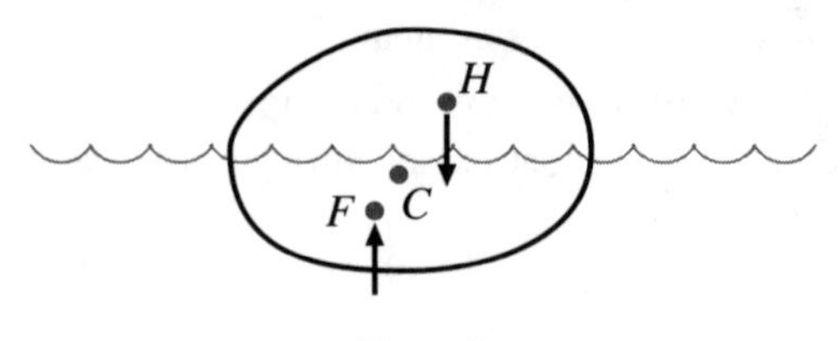

Figure 9

In these figures H happens to be to the right of F, that is, to the right of the vertical line through F. This means that the object will turn clockwise. If H were left of F, then the object would turn counterclockwise.

Since C, the center of gravity of the entire object, lies between F and H, it follows that H lies to the right of F only when C does. Similarly, H lies to the left of F only when C does. The force at H is downward. So, whether the object spins clockwise or counterclockwise depends on whether C lies right or left of F, as shown in Figure 10.

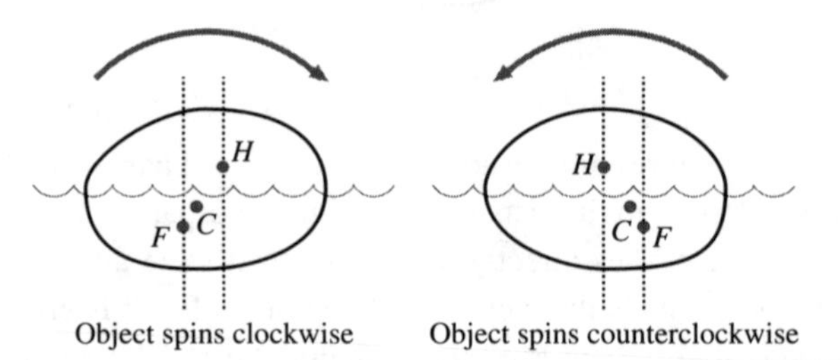

Figure 10

Though Archimedes uses the force at H when determining the direction of spin, it is more convenient to consider only the forces at C and F. At C there is the downward force of gravity acting on the entire object; at F there is the upward force of the water. In short, Archimedes' analysis of equilibrium depends on finding out whether the center of gravity of the solid (C) is left or right of the center of gravity of the submerged part (F).

A Section of a Sphere

Archimedes illustrates his general approach by considering a segment of a solid sphere, the solid mentioned at the beginning of the chapter. He asserts that if its flat base does not touch the water, the solid will move back up toward the position where its axis is perpendicular to the water surface. I will treat the case where the part removed is less than a hemisphere, that is, where it does not contain the center of the sphere.

Let us see how Archimedes reaches the conclusion just mentioned. Figure 11 shows the solid tilted from the vertical.

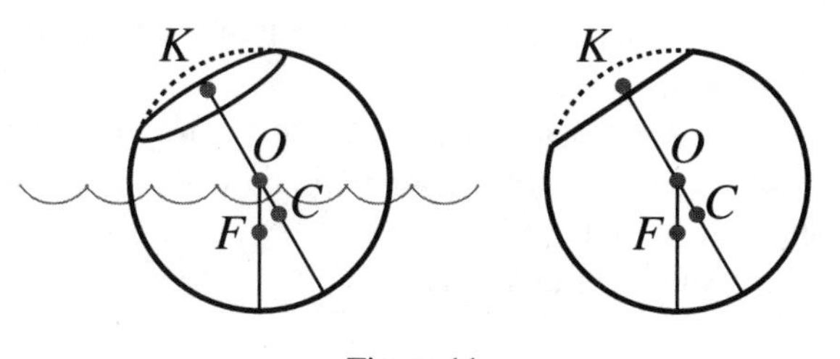

Figure 11

The center of gravity of the removed part is K. O is the center of the ball, F the center of gravity of the submerged part, and C the center of gravity of the part of the sphere remaining after the "cap" is removed. Note that O lies between C and K.

Exercise 1. Why does O lie between C and K? Why does F lie on the vertical line through O? (In fact, it lies below O, as Exercise 2 confirms, but this isn't needed.)

Since C is to the right of O it lies to the right of F, and the segment tends to turn clockwise. That means that if the segment is displaced from the vertical, it tends to right itself and return to the position where its base is parallel to the water surface. The density of the wood, hence the depth the object is submerged, plays no role. Hence this includes the case when there is no water at all and the sphere is simply placed on a flat surface.

Exercise 2. Why does F lie below O?

Exercise 3. Examine the case where the removed part contains the center of the sphere, O.

Exercise 4. Archimedes also treats the "upside-down" case, when the flat base is completely submerged. Analyze this case.

The Paraboloid Right Side Up

We now turn to a far more interesting solid, a paraboloid of revolution. Archimedes investigates nine cases, five with the base above the water, four with the base submerged. The cases depend on how much of the paraboloid is cut off and on the density of the wood, which determines how much of the segment is submerged. We will show enough of these cases to illustrate Archimedes' thinking.

In the first case, the density of the wood plays no role. The question is, "How shallow must a section of the paraboloid be so that it returns toward the vertical if displaced." It is assumed that the base of the section is completely above the water, as shown in Figure 12.

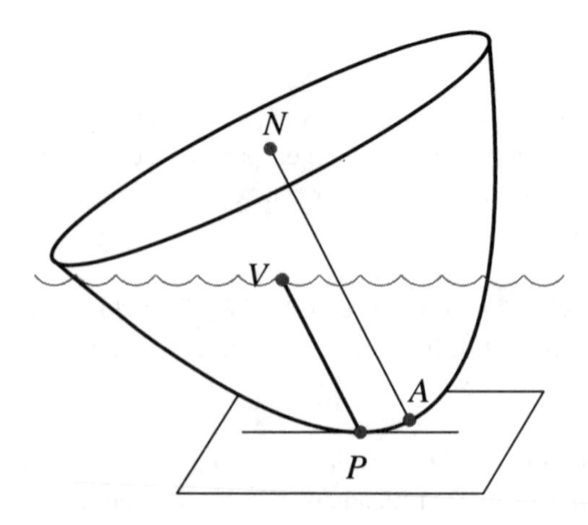

Figure 12

The vertex of the section is A, and AN is the axis of the paraboloid. P is the point of contact of the tangent plane parallel to the water surface and PV is the axis of the submerged part of the paraboloid. PV is parallel to AN.

The center of gravity, C, of the entire section is two-thirds of the way from A to N on the axis AN. (The mechanical argument for this is presented in Chapter 5.) Similarly, F, the center of gravity of the submerged part, is two-thirds of the way from P to V on the axis PV, as shown in Figure 13, a side view of the paraboloid.

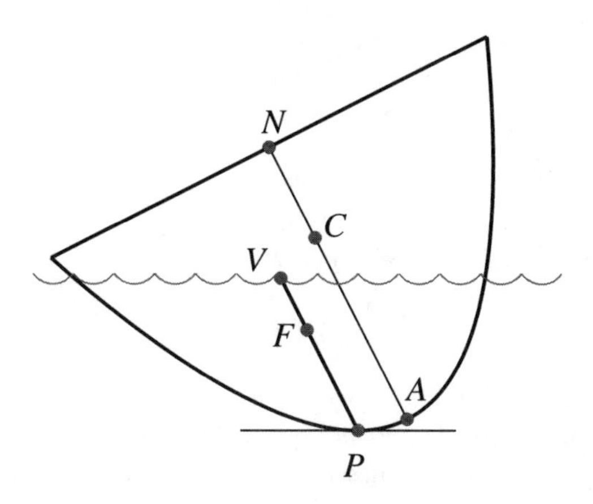

Figure 13

In Figure 13, the paraboloid is tilted counterclockwise from the vertical. Archimedes wants to know under what conditions it will tend to move clockwise, back toward the vertical. (Were the paraboloid a boat, this would certainly imply a desirable stability.) In other words, Archimedes asks, "When is C to the right of F?"

If the segment is very tall, AN will be long and C will then be high on the axis—and could easily be to the left of F. We will see how long AN can be with C still to the right of F. We will measure AN in terms of f, the distance from A to the focus.

Draw the vertical line through C. It meets PV at K. The line through K perpendicular to PV meets AN at O, as in Figure 14. Note that OC is $2f$, since triangle KOC in Figure 14 is congruent to triangle POC in Figure 5. Archimedes wants to know under what conditions K lies below F, as it happens to in the figure.

If K is lower than P on the line PV extended, then surely K would be lower than F. This will certainly happen if O is lower than A, that is, if AC is

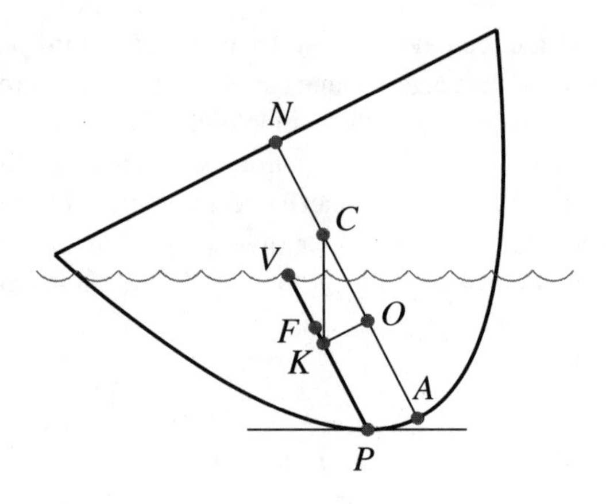

Figure 14

less than OC. Since AC is $\left(\frac{2}{3}\right) AN$ and OC is $2f$, this will occur when

$$\frac{2}{3} \cdot AN < 2f.$$

In short, *if AN is less than 3f, the section will return to the vertical, no matter how high the water is (that is, no matter what the density of the wood is).*

Figure 15 shows a section in which $AN = 3f$. It is quite shallow.

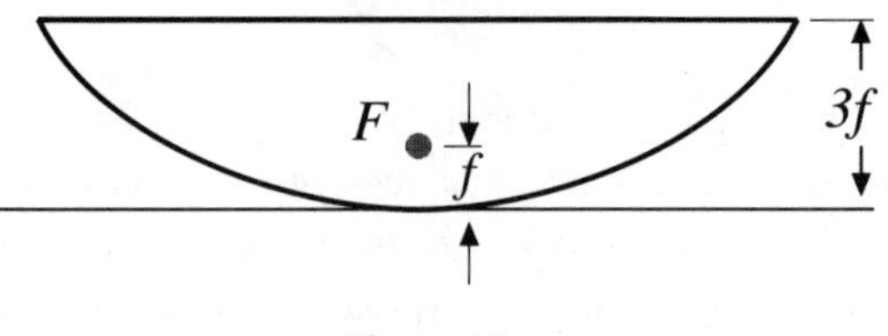

Figure 15

If the parabola has the equation $y = x^2$, its focus is at $\left(0, \frac{1}{4}\right)$ and the point the furthest right in Figure 15 is $\left(\frac{\sqrt{3}}{2}, \frac{3}{4}\right) \approx (0.87, 0.75)$. Any segment shallower than the one in Figure 15, that is, with an axis shorter than $\frac{3}{4}$, is in stable equilibrium, even if no water is present.

Exercise 5. Investigate the case when $AN = 3f$.

If the paraboloid is made of a dense wood, it will sink far and most of it may be submerged. In this case, even though the section isn't shallow, the water may offer a stabilizing force tending to right the section. Archimedes determines the relation between the height of the section and the density of the wood that still results in stable equilibrium. Let us see how he does this.

Having disposed of the case when AN is less than $3f$, he now assumes that AN is greater than $3f$. Let the density of the wood be s, which is less than 1 since the body floats. How large must s be for the water to push the body back to the position in which its axis is vertical?

In Figure 16 the axis AN is now longer than $3f$. Hence AC is longer than $2f$, which implies that O is above A. Archimedes wants to know when PK is less than PF. Instead, he determines when AO, which is larger than PK, is less than PF.

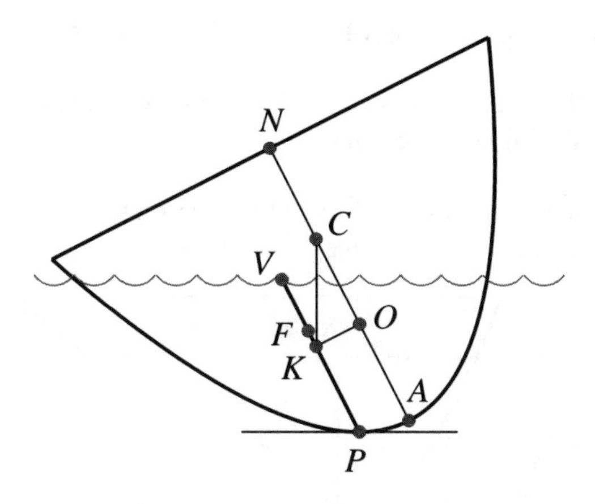

Figure 16

Now, $AO = AC - OC$; hence, $AO = \frac{2}{3} \cdot AN - 2f$. Also, $PF = \frac{2}{3} \cdot PV$. Therefore, AO being less than PF amounts to the inequality

$$\frac{2}{3}AN - 2f < \frac{2}{3}PV.$$

This simplifies to

$$AN - 3f < PV.$$

Archimedes must translate this inequality into an inequality that involves the density s. To do this, he must bring in the volumes of the submerged section and of the whole body. Since the ratio of these volumes is PV^2/AN^2, he divides by AN and squares, obtaining

$$\left(\frac{AN - 3f}{AN}\right)^2 < \frac{PV^2}{AN^2}.$$

Since the quotient on the right is the ratio of the two volumes mentioned, and this ratio is the density s of the wood, he has answered his question. If AN is greater than $3f$, and

$$s > \left(\frac{AN - 3f}{AN}\right)^2,$$

the paraboloid will return to its vertical position. In other words, its vertical position is a stable equilibrium.

As a specific example, take the case $AN = 6f$. Then for stability Archimedes wants $s > [(6f - 3f)/6f]^2 = \frac{1}{4}$.

Exercise 6. Investigate the case when $s = [(AN - 3f)/AN]^2$.

Exercise 7. If the density of the wood is 0.81, how high a section of the paraboloid is stable in its vertical position?

Exercise 8. (Not in Archimedes' work) Instead of a paraboloid of revolution, consider a prism with a parabolic base, as in Figure 17.

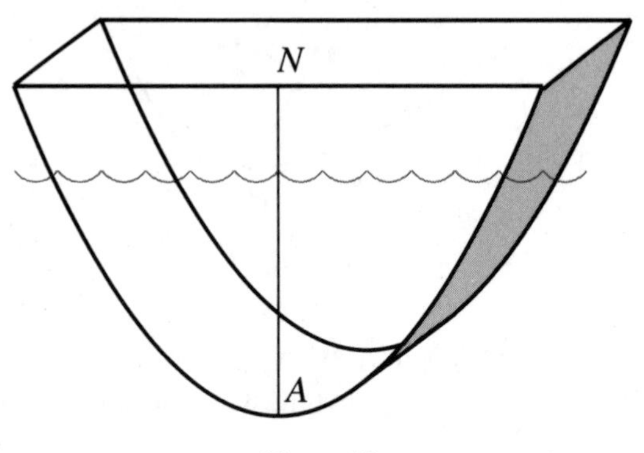

Figure 17

Investigate the stability of this boat-shaped body in the two cases analogous to the two considered for the paraboloid.(Assume that a uniform cargo fills the hull.) As shown in Chapter 7, the center of gravity of a parabola is on its axis, three-fifths of the way from its vertex A to N. The area of a parabolic section is proportional to $PV^{3/2}$, where PV is defined as for the paraboloid.

Archimedes then considers the case when the base is still above the water but its edge just touches the water's surface, as in Figure 18.

Assuming that $3f < AN < \left(\frac{15}{2}\right) f$, he shows that the paraboloid does not stay in this position. (The details of this case are deferred to Appendix B, which

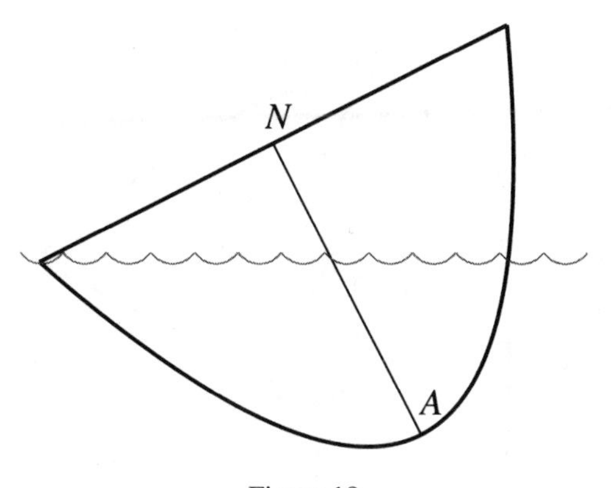

Figure 18

shows why the number $\frac{15}{2}$ suddenly appears.) Rather, there is a force pushing it toward the upright position. This leaves the possibility that there may be a position of equilibrium somewhere between the position in Figure 18 and the upright. He then examines these positions in two propositions.

In the first proposition, in which Archimedes treats the case $3f < AN < \left(\frac{15}{2}\right)f$, the density of the wood is less than $(AN - 3f)^2/AN^2$. He shows that if the solid is placed with its base above the water and tilted, it will not return to the vertical. It will not remain in any position except one that makes a certain angle with the vertical. This angle depends on AN and the density of the wood.

Finally, Archimedes examines the situation when $AN > \left(\frac{15}{2}\right)f$. This case breaks into five special ones, depending on the density of the wood. For each one, he determines the position of rest. For the details of these cases, which we omit, see either the Heath or Dijksterhuis translations.

The Paraboloid Upside Down

Archimedes also investigates a capsized paraboloid, one whose base is totally submerged, as in Figure 19.

The center of gravity of the submerged part is H and the center of gravity of the exposed part is F. C is the center of gravity of the whole segment. In this case, if H, hence C, is to the left of F, the paraboloid tends to return to the vertical. This is precisely the geometry of the first case of the paraboloid right side up. (Just turn the page upside down.) *Therefore, when AN is less than $3f$, the capsized paraboloid returns to the vertical.*

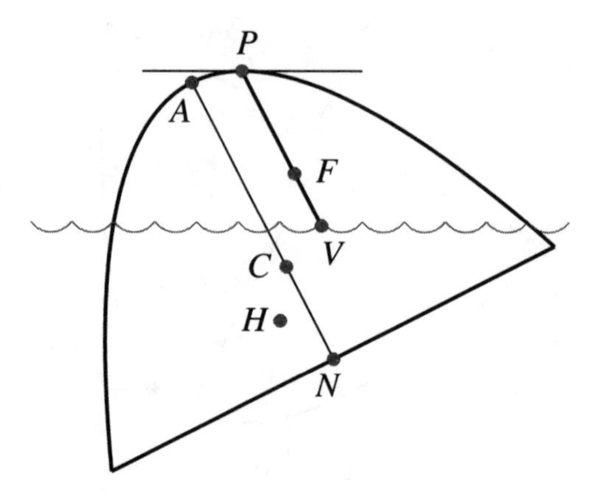

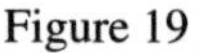
Figure 19

Next Archimedes considers the capsized paraboloid for which AN is greater than $3f$. His analysis is like that for the second case of the paraboloid with an exposed base, and he concludes that if AN is greater than $3f$ and the density of the wood is less than

$$\frac{AN^2 - (AN - 3f)^2}{AN^2},$$

then the upside-down paraboloid returns to the vertical.

Exercise 9. Examine the case just mentioned.

Archimedes also investigates the case where the submerged base just touches the surface of the water and the case where $3f < AN < \left(\frac{15}{2}\right)f$ and the density of the wood is greater than the density mentioned just before Exercise 9.

In this study of hydrostatics, Archimedes demonstrates not only his geometric and physical insight but also his thoroughness and persistence, essential components of a successful scientist's character. Though we can grasp all of these discoveries in a day or two, I suspect it took him months to carry out his analysis. For him, this research, which sounds so practical, was a study in mathematics, as pure and abstract as geometry. There is no evidence that it was applied to the design of ships in ancient times.

9

The Spiral

Often a mathematical concept is given the name of a mathematician who had nothing to do with it. The Archimedean Spiral, however, is properly named, for Archimedes seems to be the first person to have defined it. He introduces it in his work *On Spirals* in these words,

> A straight line in the plane revolves at a constant rate about one of its ends (which remains fixed) and returns to its starting position. At the same time a point moves at a constant rate along the moving line beginning at the fixed end and describes a spiral.

The spiral is shown in Figure 1, where the fixed end is labeled O, for "origin."

In this chapter we will see how Archimedes finds the area bounded by the spiral and the initial line, shaded in Figure 1. It turns out that his technique would also enable him to find the volume of a sphere. (In the next chapter he finds that volume by a different method.)

But first I quote more from his introduction to *On Spirals*, for it offers a glimpse of his attitude toward mathematics and mathematicians:

> Archimedes to Dositheus greeting:
>
> Of most of the theorems I sent to Conon, and of which you ask me from time to time to send you the proofs, the demonstrations are already before you in the books brought to you by Heracleides. Some more are also in what I now send you. Do not be surprised at my taking a considerable time before publishing these proofs. This has been owing to my desire to

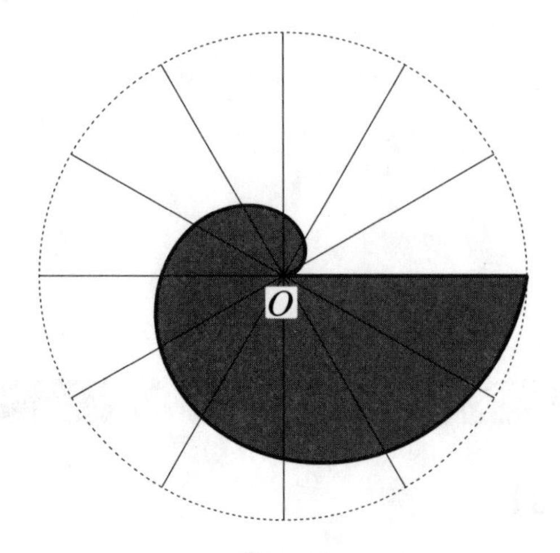

Figure 1

> communicate them first to persons engaged in mathematical studies and anxious to investigate them.
>
> How many theorems in geometry which seemed at first impossible are in time successfully worked out! Now Conon died before he had sufficient time to investigate the theorems referred to. Otherwise he would have discovered all these things, and would have enriched geometry by many other discoveries besides. For I well know that it was no common ability that he brought to bear on mathematics, and that his industry was extraordinary.
>
> Though many years have elapsed since Conon's death, none of the problems has been stirred by a single person. I wish to review them, particularly since two of them are impossible and may serve as a warning to those who claim to discover everything but who produce no proofs. They may be shown as having pretended to discover the impossible.

Archimedes then lists seven problems and continues:

> Heracleides brought you the proofs of all these propositions. The proposition stated next after these was wrong: If a sphere is cut by a plane into two unequal parts, then their volumes are proportional to the squares of their surface areas. That this is wrong is obvious by what I sent you before. For it included this proposition: If a plane perpendicular to a diameter cuts a sphere into two parts, their surface areas are proportional to the lengths cut off the diameter.

Exercise 1. Why does Archimedes' (true) theorem show that his alleged proposition was wrong?

He then introduces his spiral.

To find the area within the spiral he uses the following two tools, one geometric, one arithmetic.

Geometric: The area of a circle is proportional to the square of its radius. In other words, if one circle has area A and radius a, and another circle has area B and radius b, then

$$\frac{A}{B} = \frac{a^2}{b^2}.$$

This had been proved in Book 12 of Euclid.

Arithmetic: The following inequalities on sums of squares hold:

$$1^2 + 2^2 + \cdots + (n-1)^2 < \frac{n^3}{3} < 1^2 + 2^2 + \cdots + (n-1)^2 + n^2.$$

His proof, which appeared in *On Spirals*, is to be found in Chapter 6.

To begin the calculation of the area within the spiral, let the revolving straight line have length a. Its moving end sweeps out a circle of radius a and area A, shown in Figure 1. Archimedes proves that *the area bounded by the spiral and initial line is one-third the area of the circle.*

Archimedes begins by cutting the circle into n congruent sectors by rays starting at O, as in Figure 2, which shows the case $n = 12$.

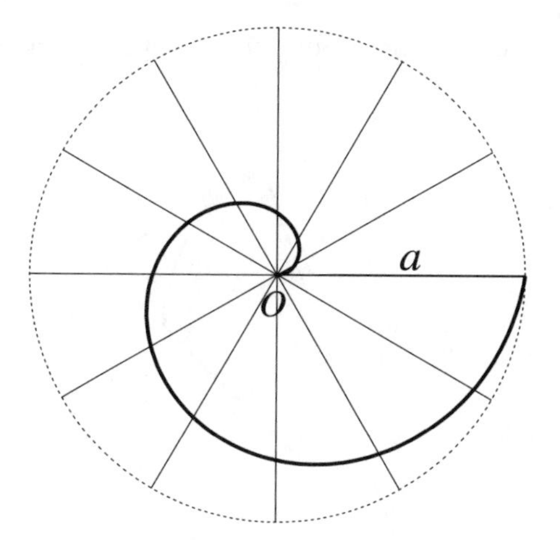

Figure 2

Each sector has area A/n.

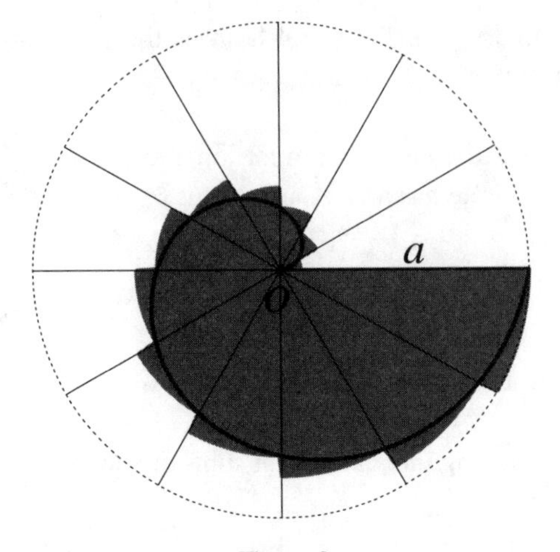

Figure 3

Archimedes then overestimates the area of the spiral by n "special" sectors of smaller circles, as shown in Figure 3.

Each special sector is bounded by the spiral and two adjacent rays. Its radius is the longest distance from O to that part of the spiral that lies in the special sector. The smallest of the n special sectors has radius a/n. The radius of the largest is a.

The typical special sector, say the kth, is shown in Figure 4.

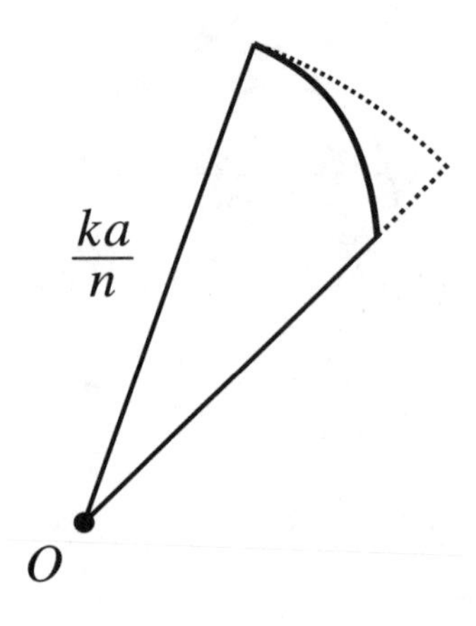

Figure 4

Since the point on the spiral moves out at a constant speed as the ray turns, the radius of the kth sector is ka/n. Archimedes now calculates the total area of the n special sectors, expressing the sum in terms of the circle whose area is A.

Let A_k be the area of a circle of radius ka/n. Then

$$\frac{A_k}{A} = \frac{\left(\frac{ka}{n}\right)^2}{a^2} = \frac{k^2}{n^2}.$$

Thus

$$A_k = \frac{k^2}{n^2}A.$$

Since the special sector shown in Figure 4 is only an nth of the circle of radius ka/n, Archimedes has

$$\text{area of } k\text{th special sector} = \frac{1}{n} \cdot \frac{k^2}{n^2}A = \frac{k^2}{n^3}A.$$

The total area of all these n special sectors is therefore

$$\frac{1^2 + 2^2 + \cdots + n^2}{n^3}A.$$

But Archimedes has shown that the sum in the numerator is greater than $n^3/3$. Therefore the overestimate of the area of the spiral is larger than $A/3$.

Archimedes obtains an underestimate of the area of the spiral in a similar manner. The only difference is that the kth special sector lies inside the spiral. Its radius is the shortest distance from O to the part of the spiral that lies in the given sector of the circle of radius a. This typical underestimate is shaded in Figure 5. The kth underestimate has radius $(k-1)/n$.

The smallest such special sector, which occurs when k is 1, has radius 0, and is invisible to the naked eye. The largest, corresponding to $k = n$, has radius $(n-1)a/n$. The total area of these n underestimates is

$$\frac{1^2 + 2^2 + \cdots + (n-1)^2}{n^3}A.$$

Since Archimedes already has shown that the sum in the numerator is less than $n^3/3$, he now has

$$\text{underestimate} < \frac{A}{3} < \text{overestimate}.$$

At this point there are two ways to proceed: the modern and the ancient Greek. First let's use the modern approach.

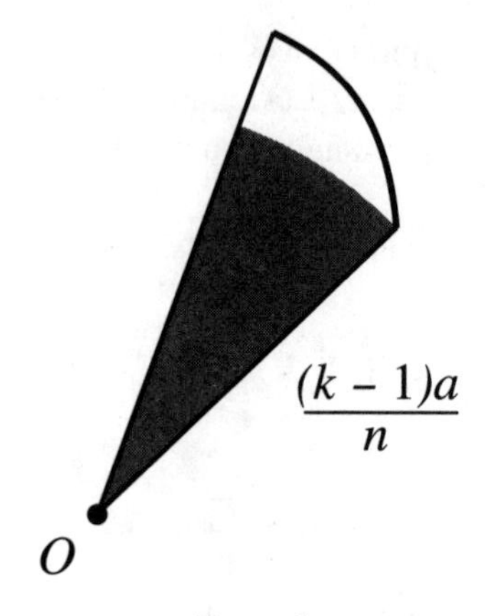

Figure 5

Letting S be the area of the spiral, we have

$$\text{underestimate} < S < \text{overestimate}.$$

That means each of the fixed numbers, $A/3$ and S, is trapped between every underestimate and every overestimate. But the difference between the two estimates for a given n is just

$$\frac{n^2}{n^3}A = \frac{A}{n}.$$

Since we can make A/n as small as we please by choosing n large, the difference between the under- and overestimates can be made as small as we please. Therefore S and $A/3$, which lie between these two estimates, must be the same number. We conclude that the area within the spiral is exactly one-third the area of the circumscribing circle.

Archimedes follows a different but logically equivalent route, in keeping with the style of his times. He shows that each of the assumptions, "S greater than $A/3$" and "S less than $A/3$," leads to a contradiction.

Take the first case, where he assumes that the area within the spiral is greater than $A/3$, hence $S - A/3 > 0$. Pick n so large that an underestimate I differs from S by less than $S - A/3$. (This is possible because an underestimate differs from the corresponding overestimate by A/n.) Thus Archimedes has $S - I < S - A/3$; hence, I is greater than $A/3$. This contradicts the fact that each underestimate is less than $A/3$.

Exercise 2. Use a similar approach to show that S is not less than $A/3$.

It follows that S must be $A/3$.

Using the same approach, Archimedes goes on to find the area within the spiral after several turns. The details are similar to those for the case of a single turn.

Excursions

The inequalities

$$1^2 + 2^2 + \cdots + (n-1)^2 < \frac{n^3}{3} < 1^2 + 2^2 + \cdots + (n-1)^2 + n^2$$

permit us to find other areas and volumes, as the remaining exercises illustrate.

Exercise 3.

(a) Using over- and underestimates, find the area under $y = x^2$ and above the interval on the x-axis whose ends are $(0, 0)$ and $(a, 0)$, where a is positive. (To construct the estimates, divide the interval of length a into n equal segments by $n - 1$ points and use lines through them parallel to the y-axis to divide the region into n narrow regions that can be approximated by rectangles of the same width, a/n.)

(b) Use the result in (a) to find the area between the parabola and the line through $(-a, a^2)$ and (a, a^2)

Exercise 4. Find the volume of the pyramid in Figure 6. Its base is a square of side a and its height is h.

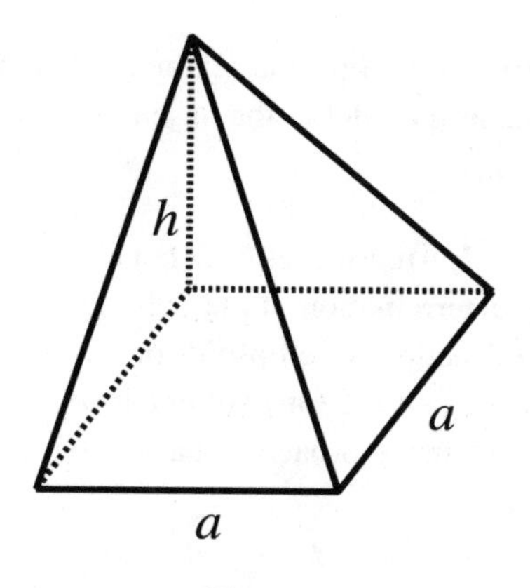

Figure 6

To estimate the volume, divide the height into n sections, each of length h/n, by $n-1$ points. Through each of these points there is a plane parallel to the base. Estimate the volume that lies between any two consecutive planes.

Exercise 5. Like Exercise 4, only the solid is a cone of height h and with a circular base of radius a.

Exercise 6. Let R be a region with area A in the plane. A point P is at a distance h from the plane. A solid is made up of all the line segments that have one end at P and the other end in R, as shown in Figure 7. Using over- and underestimates in the style of Archimedes, show that its volume is $hA/3$.

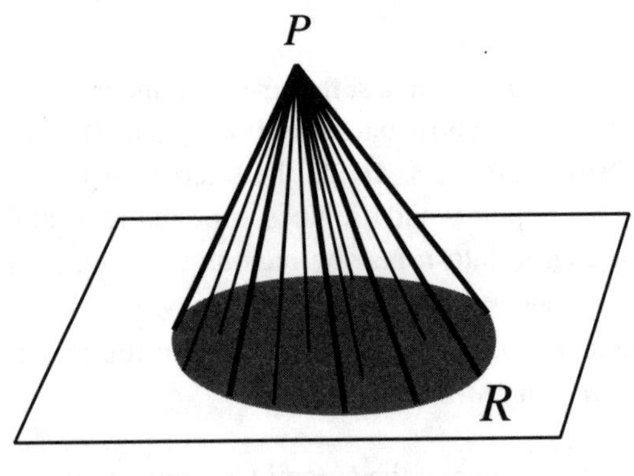

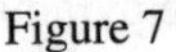
Figure 7

Exercise 7. In the same style, find the volume of a hemisphere of radius a. Use equally spaced planes parallel to the base to cut the hemisphere into thin sections resembling coins.

In spite of Exercise 7, Archimedes does not use the sum of squares in what was probably his first determination of the volume of a sphere. However, in a work on the volumes of sections of ellipsoids of revolution he uses approximations by parallel thin disks, whose total volume involves a sum of squares. The next chapter presents his first approach and also the method by which he finds the surface area of a sphere.

10

The Sphere

As we mentioned in Chapter 1, Archimedes was so proud of his discovery of the surface area and volume of a sphere that he wanted it memorialized on his gravestone. As he summarized his find,

> Every cylinder whose base is the greatest circle in a sphere and whose height is equal to the diameter of the sphere has a volume equal to $\frac{3}{2}$ the volume of the sphere. Its surface, including the bases, is $\frac{3}{2}$ the surface of the sphere.

He went on,

> Now these properties were all along naturally inherent in the figures, but remained unknown to those who were before my time engaged in the study of geometry. Having, however, discovered these properties, I have no hesitation in setting them side by side both with my earlier investigations and with those theorems of Eudoxus which are certainly well established. Though these discoveries were naturally inherent in the figures all along, yet they were in fact unknown to the many able geometers who lived before Eudoxus, and had not been observed by anyone.
>
> Now it will be open to those who possess the requisite ability to examine these discoveries of mine. They ought to have been published while Conon was still alive, for I am sure that he would best have been able to grasp them and to pronounce upon them the appropriate verdict. But, as I judge it well to communicate them to those who are conversant with mathematics, I send them to you with the proofs, which it will be open to mathematicians to examine. Farewell.

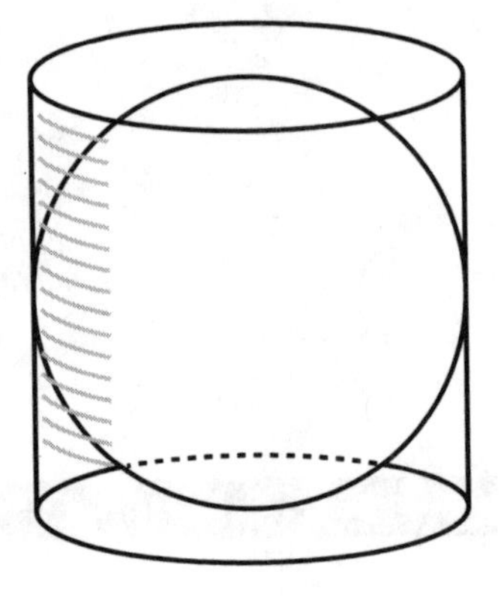

Figure 1

The cylinder that he mentions circumscribes the sphere as shown in Figure 1.

That the ratio is the same, namely $\frac{3}{2}$, for both the area and the volume is not a coincidence, as the next exercise shows.

Exercise 1. Let P be a polyhedron circumscribing a sphere B of radius r. Let the surface area of the sphere be $S(B)$ and the surface area of the polyhedron be $S(P)$.

(a) Show that the volume of P is $(r/3)S(P)$.

(b) Show that volume of P/volume of $B = S(P)/S(B)$.

His Method

Archimedes finds the surface area of a sphere in the same way that he dealt with the area within a spiral: by examining under- and overestimates that he could calculate. He begins by inscribing a regular polygon with an even number of sides in a circle. Then he spins these polygons around a diagonal through opposite vertices to produce a surface that approximates the surface of the sphere formed by spinning the circle around the same diameter. Each edge of the polygon sweeps out a band, as shown shaded in Figure 2.

After computing the area of each band, Archimedes adds their areas to estimate the area of the spherical surface.

However, there is a troublesome issue here that does not arise in the case of the spiral. The sectors situated inside the spiral clearly have a smaller area

Figure 2

than the area of the region within the spiral. But is the area of Archimedes' approximation of the surface area of a sphere necessarily smaller than that area? Surely the *volume* bounded by the approximating surface is less than the *volume* of the sphere. That doesn't imply a similar inequality for surface *areas*. A very bumpy surface lying in a solid bounded by a very smooth surface can have a much larger area than the outer surface. Similarly, a very tortuous curve can be very long, yet easily fit within a very short curve, as in Figure 3.

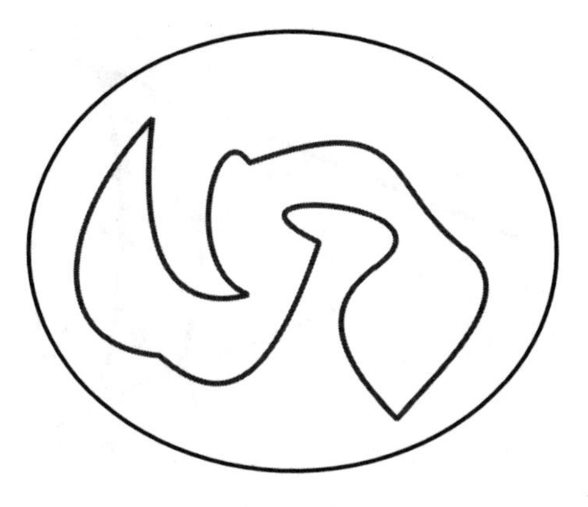

Figure 3

Archimedes, seeing that this is a critical consideration, makes an explicit assumption that involves the notion of convexity, which we introduced in Chapter 3.

A region in the plane or in space is convex if for any pair of points in the region the entire line segment joining them also lies in the region. A curve in the plane is called convex if it is the boundary of a convex region. Similarly, a surface is called convex if it bounds a convex region in space.

Archimedes assumes without proof that a convex curve situated in the plane region bounded by another curve is shorter than that curve. This assumption will be needed in the next chapter when Archimedes estimates the circumference of a circle. Similarly, he assumes that a convex surface situated in the region bounded by another surface has a smaller area than that of the outer surface. He makes use of this assumption in the present chapter. These two assumptions became theorems in the theory of convexity developed in the nineteenth and twentieth centuries.

Exercise 2. Using elementary geometry, show that a convex polygon situated in the region bounded by another polygon is shorter than the outer polygon.

The Area of a Cone

Archimedes views a band as the difference of two cones, as in Figure 4.

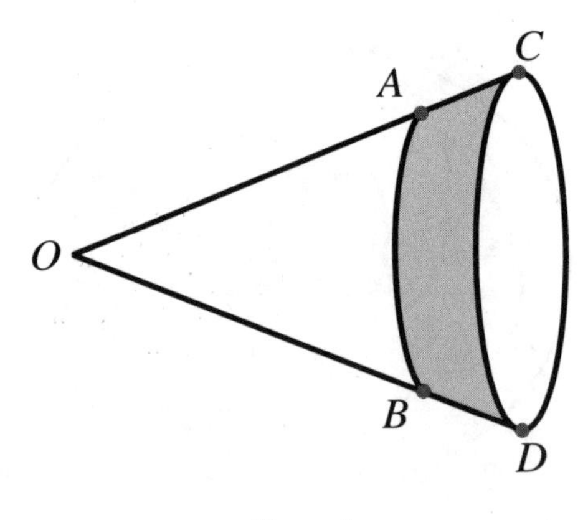

Figure 4

The band, shaded in Figure 4, is obtained by removing cone OAB from cone OCD. Therefore, Archimedes first determines the area of the curved surface of a cone (excluding its base). He does this by approximating the base by a polygon and using the edges of the polygon as the bases of triangles whose opposite vertex is the tip of the cone, as indicated in Figure 5.

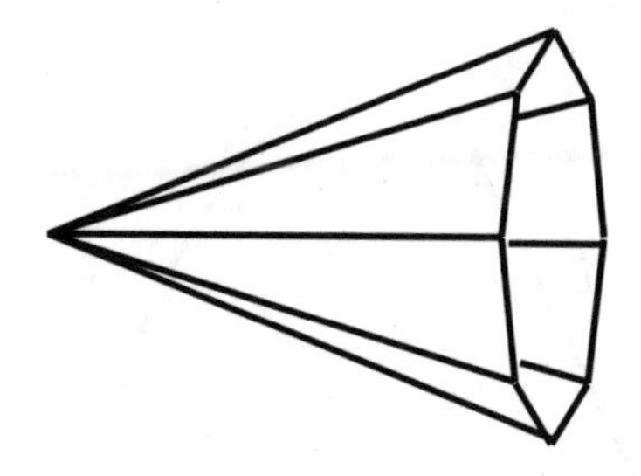

Figure 5

Going through the many details of this approach would take us away from our main interest, finding the surface area of a sphere. Instead, we will use an informal approach described by Archimedes: view the surface of the cone as a "triangle":

Let the cone have slant height l and radius r, as in Figure 6.

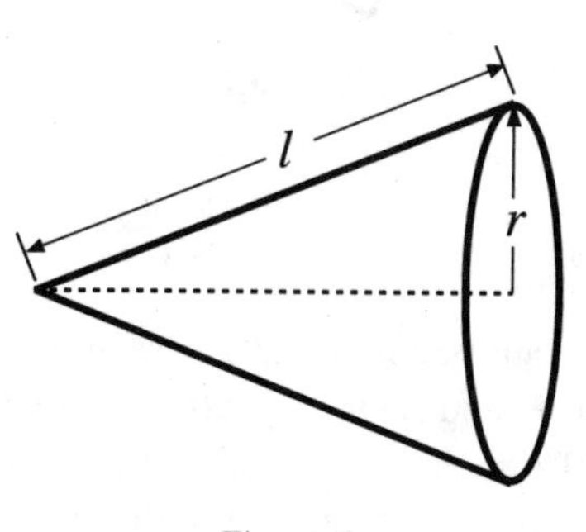

Figure 6

Make a straight cut from the vertex to the base and lay the curved surface flat, as in Figure 7.

We may think of the flat region in Figure 7 as a "triangle" of height l and base $2\pi r$. Thus its area is presumably

$$\frac{1}{2} \cdot l \cdot 2\pi r = \pi r l.$$

Exercise 3. How would you use Archimedes' approximation by the triangles shown in Figure 5 to obtain the same formula?

With the aid of this formula, Archimedes determines the area of a band.

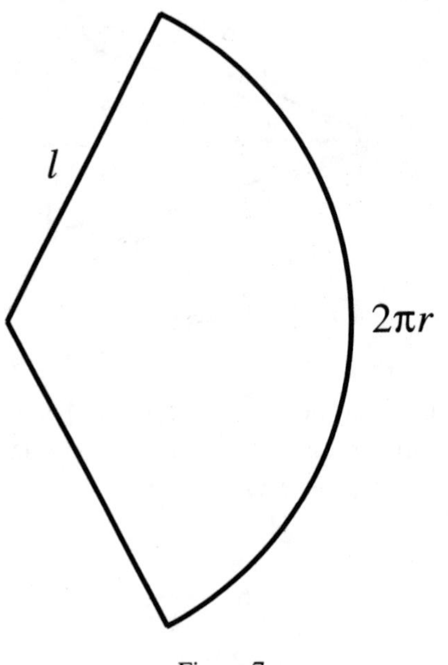

Figure 7

The Area of a Band

The typical band has slant height s and is the difference of two cones. The smaller cone has radius r_1 and slant height l_1 , while the larger cone has radius r_2 and slant height l_2. These dimensions are shown in Figure 8.

The area of the band is the difference of the areas of the two cones,

$$\pi r_2 l_2 - \pi r_1 l_1.$$

But we wish to express this in terms of the two radii and the slant height, s, of the band.

We know that

$$l_2 = l_1 + s$$

and, by similar triangles,

$$\frac{r_2}{r_1} = \frac{l_2}{l_1}.$$

This last equation implies that $r_2 l_1 = r_1 l_2$.

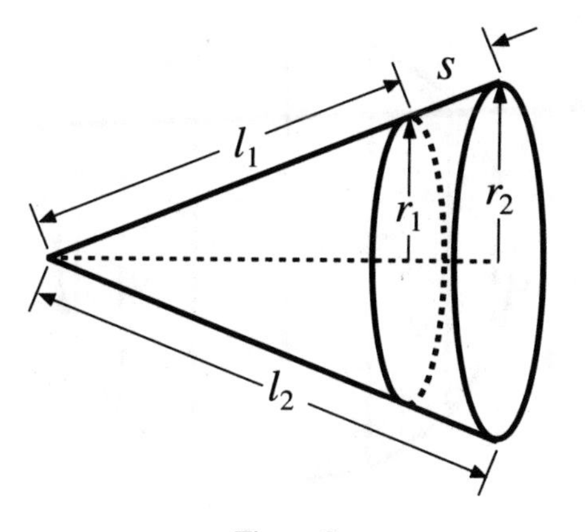

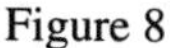
Figure 8

Using these equations, we can now express the area of the band in terms of the radii and s:

$$\begin{aligned}
\pi(r_2 l_2 - r_1 l_1) &= \pi(r_2(l_1 + s) - r_1 l_1)\\
&= \pi(r_2 l_1 + r_2 s - r_1 l_1)\\
&= \pi(r_1 l_2 + r_2 s - r_1 l_1)\\
&= \pi(r_1(l_2 - l_1) + r_2 s)\\
&= \pi(r_1 s + r_2 s)\\
&= \pi(r_1 + r_2)s.
\end{aligned}$$

In short, the area of the band is $\pi(r_1 + r_2)s$. Now that we have the formula for the area of a band, we can join Archimedes as he estimates the surface area of a sphere.

The Surface Area of a Sphere

The surface of a sphere of radius r is obtained by spinning a circle of radius r around a diameter. Archimedes approximates this circle by an inscribed regular polygon of $2n$ sides. Then he spins the polygon around a diameter that joins opposite vertices. This produces a surface formed of n bands. Figure 9 shows the case $n = 8$.

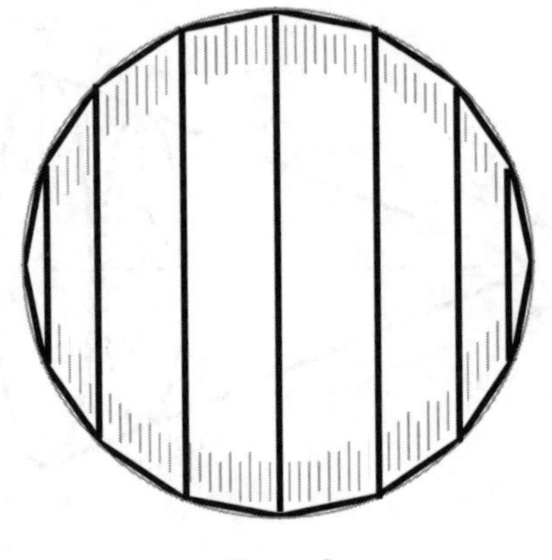

Figure 9

He also will construct in a similar fashion an overestimate of the surface area of the sphere. Then he examines how these estimates behave as n increases.

Consider the underestimates. The n bands are cut out by $n-1$ planes that are at right angles to the diameter around which the polygon has been spun. The one furthest to the left is a cone of radius r_1, while the one furthest to the right is a cone of radius r_{n-1}. The left-most radii of the other $n-2$ bands are $r_2, r_3, \ldots, r_{n-2}$, as shown in Figure 10, which shows a cross section of the approximating surface by a plane through a diameter. The edges of the polygon all have length s.

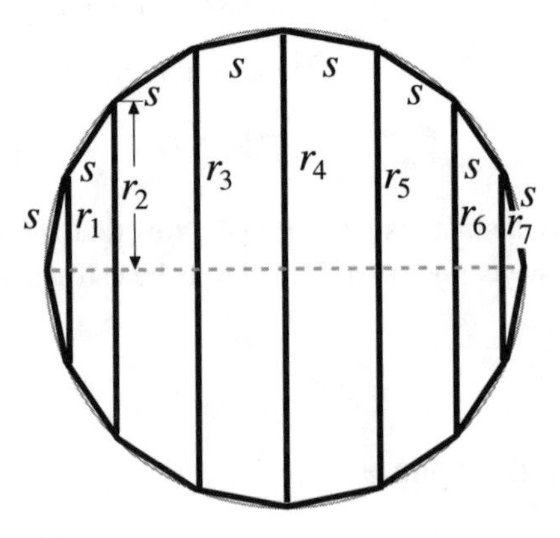

Figure 10

Since each side of the polygon has length s, the total area of the n bands is

$$\pi r_1 s + \pi(r_1 + r_2)s + \pi(r_2 + r_3)s + \cdots + \pi(r_{n-2} + r_{n-1})s + \pi r_{n-1} s,$$

which equals

$$2\pi(r_1 + r_2 + r_3 + \cdots : +r_{n-1})s.$$

Imagine Archimedes as he faced the sum of the radii in the last expression. I wonder how long it took him to come up with the following way to deal with what must have appeared to him to be a most unusual and difficult sum.

This is his response. In each band (other than the two little end cones) he constructs a line that goes from its top right point to its bottom left point. These lines, together with the lines determined by the planes bounding the bands, and the altitudes a_1 and a_{n-1} of the two little cones at the ends of the diameter cut the diameter in Figure 10 into $2n - 2$ small pieces of lengths $a_1, a_2, \ldots, a_{n-1}$, each appearing twice, as shown in Figure 11.

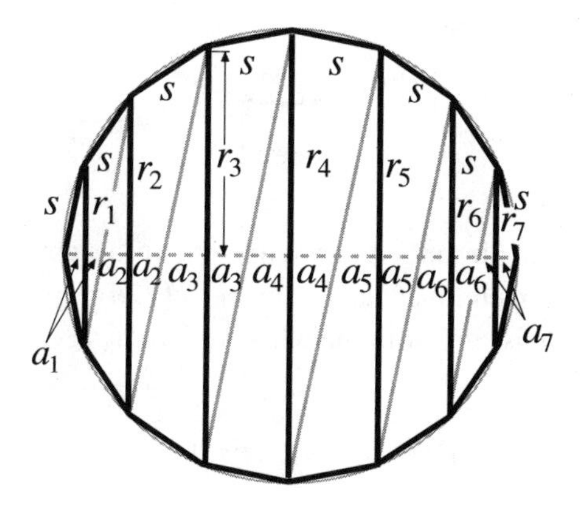

Figure 11

Exercise 4. Explain why the diameter is divided into segments that come in pairs of equal length.

Now comes the key step: All the right triangles that look similar indeed are similar. In each one the angle that has a vertex on the circle (not the angle along the diameter) subtends an arc that is $(1/2n)$th of the circle. Since such an inscribed angle equals half the angle subtended at the center of the circle by the intercepted arc, all those angles are equal. That is why the right triangles above (and below) the diameter are similar.

Since corresponding parts of similar figures are proportional,

$$\frac{r_1}{a_1} = \frac{r_2}{a_2} = \cdots = \frac{r_{n-1}}{a_{n-1}}.$$

To express this fixed ratio in terms of the circle, Archimedes draws the line joining the right-hand point of the diameter AB to the top, C, of the left-most triangle, as in Figure 12.

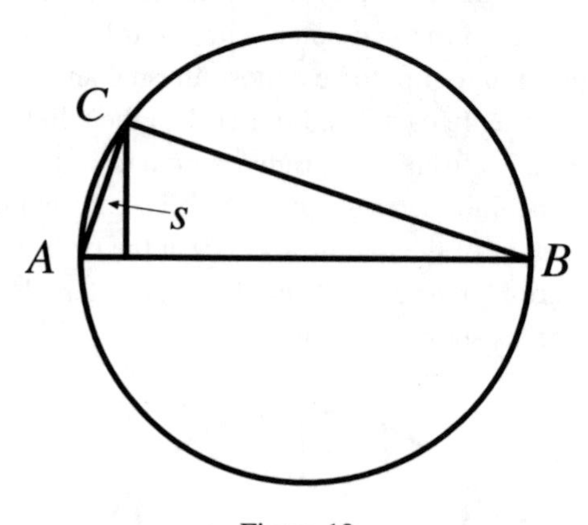

Figure 12

Angle ACB, being inscribed in a semicircle, is a right angle. Angle CBA, like the other angles mentioned, also subtends $(1/2n)$th of the circumference. So this large triangle is similar to the other right triangles. As a result,

$$\frac{r_1}{a_1} = \frac{r_2}{a_2} = \cdots = \frac{r_{n-1}}{a_{n-1}} = \frac{BC}{s}.$$

Therefore,

$$r_1 = \frac{BC}{s}a_1, r_2 = \frac{BC}{s}a_2, \ldots, r_{n-1} = \frac{BC}{s}a_{n-1}.$$

Consequently,

$$r_1 + r_2 + \cdots + r_{n-1} = \frac{BC}{s}(a_1 + a_2 + \cdots + a_{n-1}).$$

Thus the approximating area is

$$\begin{aligned} 2\pi(r_1 + r_2 + \cdots + r_{n-1})s &= 2\pi\frac{BC}{s}(a_1 + a_2 + \cdots + a_{n-1})s \\ &= 2\pi(a_1 + a_2 + \cdots + a_{n-1})BC. \end{aligned}$$

But twice the sum of all the a_k's is exactly the diameter AB:

$$2(a_1 + a_2 + \cdots + a_{n-1}) = AB.$$

Thus the lower estimate reduces, after all this, simply to

$$\pi \cdot AB \cdot BC.$$

If the sphere has radius r, then $AB = 2r$ and BC is less than $2r$.

Therefore the inner estimate is less than $\pi(2r)(2r) = 4\pi r^2$. When n is large, BC is near AB. Therefore these estimates can be arbitrarily close to $4\pi r^2$.

Even so, Archimedes does not conclude that the surface area of a sphere is $4\pi r^2$. All that he knows for sure is that the surface area is at least $4\pi r^2$.

Overestimates

To obtain estimates that are too large, he encloses the circle in a circumscribed regular polygon with $2n$ sides and carries out a similar computation. Figure 13 shows the key part of the related diagram, compared to the related portion of Figure 12.

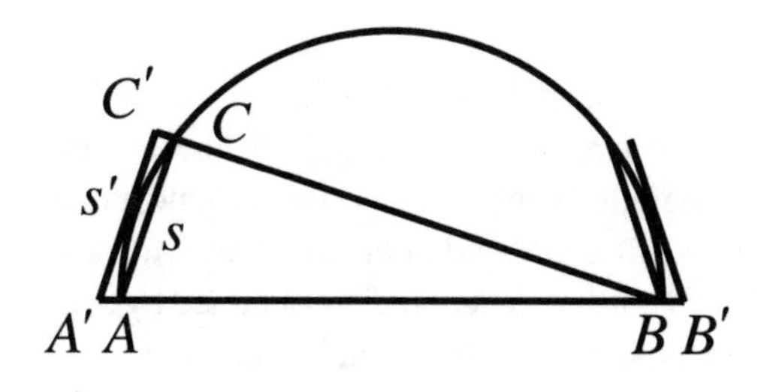

Figure 13

The circumscribing polygon, each of whose edges has length s', is the inscribed polygon of a slightly larger circle, one with diameter $A'B'$. Its area is therefore $\pi A'B' \cdot B'C'$. As n increases, both $A'B'$ and $B'C'$ approach $2r$. Hence, these overestimates approach $4\pi r^2$ as n increases. Therefore, the surface area of the sphere is at most $4\pi r^2$. It follows that its surface area is exactly $4\pi r^2$.

(Archimedes uses a proof by contradiction in the style of the one he used when finding the area within a spiral.)

Exercise 5. Using the same approach, by approximations, examine the surface area of that part of a sphere of radius r that lies between two parallel planes that meet the sphere.

The Volume of a Sphere

Archimedes finds the volume of a sphere the same way, using under- and over-estimates derived from approximating a circle by regular polygons. However, he doesn't do it in the way I expected. I thought he would approximate the sphere by parallel cylinders or by solid bands, as in Figure 14.

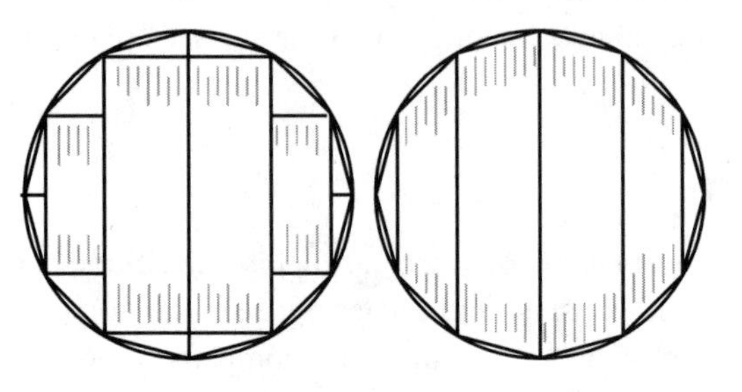

Figure 14

Instead, he views the polygon as being composed of triangles with a common vertex at the center of the circle. He then determines the volume swept out by each triangle as the circle is spun around a diameter.

In another work, *Conoids and Spheroids*, Archimedes does use approximations by parallel cylinders. A "conoid" is obtained by spinning a parabola or hyperbola about an axis, while a "spheroid" is obtained by spinning an ellipse about an axis. Among the volumes he determines is that of the part of a spheroid cut off by a plane. Since a sphere is a special case of a spheroid, this gives him a second way to obtain the volume of a sphere, expressing it in terms of the volume of a certain inscribed cone. His other approach, described in the next few pages, expresses the volume in terms of the surface area of the sphere.

Let BC be a typical edge of the rotated polygon and O the center of the sphere. The typical triangle, OBC, is shown in Figure 15. The altitude from O to the side BC has length p. (The solid swept out has no name and is hard to draw. When BC is short, it looks like the curved surface of a cone.)

However, Archimedes, who, as we saw in Chapter 5, thinks of a sphere as a cone with its spherical surface as base and the center of the sphere as its vertex, can easily predict the volume of the solid swept out by triangle OBC. This solid can be pictured as a pyramid whose base is the surface swept out by the edge BC and whose height is the altitude p. Its volume is therefore probably

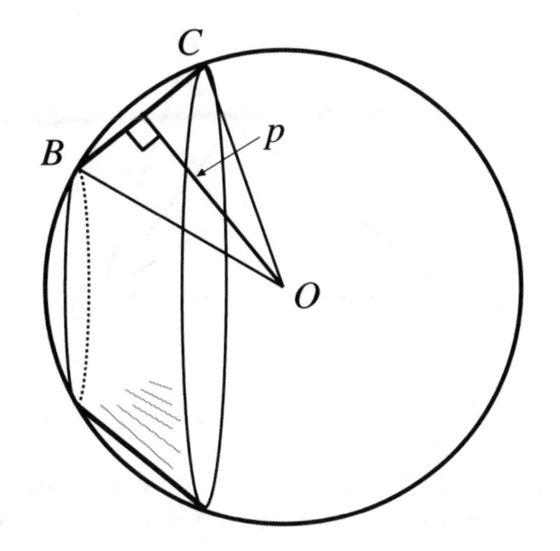

Figure 15

$$\frac{1}{3} \cdot p \cdot \text{area of surface swept out by } BC.$$

The sum of these estimates for all the edges of the polygon is then

$$\frac{1}{3} \cdot p \cdot \text{area of surface swept out by the polygon.}$$

As the number of sides of the polygon increases, p approaches the radius of the sphere, r, and the area of the surface swept out by the polygon approaches the surface area of the sphere. Therefore, it seems reasonable to expect that the volume of the sphere is

$$\frac{1}{3} \cdot r \cdot \text{surface area of sphere.}$$

All that remains for Archimedes to do is to find rigorously the volume swept out by triangle OBC.

He begins by extending BC to meet the line around which the polygon is spun. (In order that BC not be parallel to that line, choose n to be even; the number of sides of the polygon, $2n$, is then a multiple of 4.) Let D be the point where the two lines meet, as shown in Figure 16.

He views triangle OBC as the difference between the large triangles OCD and OBD. Each of these large triangles, when spun around OD, sweeps out a double cone. The one swept out by OCD is shown in Figure 17; the one swept

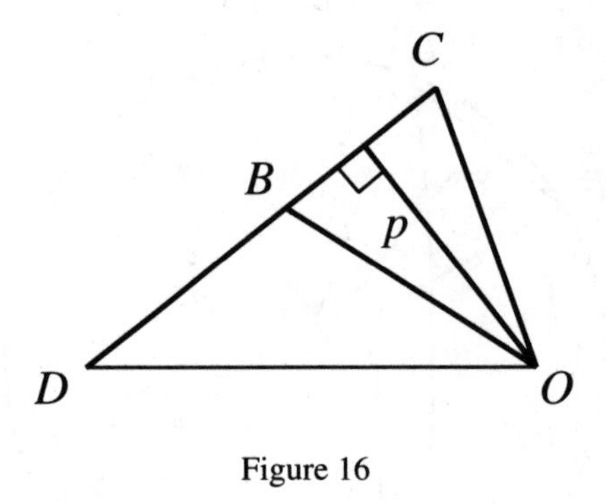

Figure 16

out by OBD is smaller. Since the volume swept out by triangle OBC is the difference of the volumes of those two double cones, all that Archimedes has to do is find the volume of a double cone.

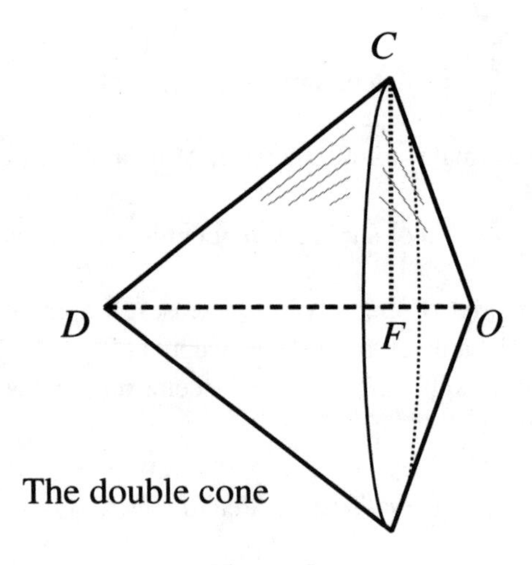

Figure 17

The curved surface of the left cone in Figure 17 has area $\pi \cdot CD \cdot CF$, while its base has area $\pi \cdot CF \cdot CF$. Therefore,

$$\frac{\text{area of curved surface}}{\text{area of base}} = \frac{CD}{CF}$$

Now, CD/CF is related to p, as similar triangles in Figure 18 show:

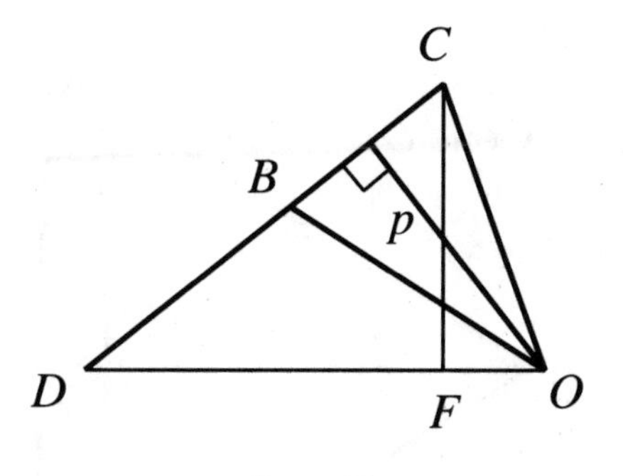

Figure 18

$$\frac{CD}{CF} = \frac{OD}{p}.$$

This will be needed in a moment.

To find the volume of a double cone, Archimedes changes its shape in two steps.

Step 1. The double cone in Figure 17 is composed of two cones that share a base of area A. Let their heights be h and h'. The sum of their volumes is

$$\frac{1}{3} \cdot hA + \frac{1}{3} \cdot h'A = \frac{1}{3} \cdot (h + h')A = \frac{1}{3} \cdot OD \cdot A.$$

Thus the double cone has the same volume as the cone that has height OD and the same base as each of the cones that form the double cone. This cone is shown in Figure 19. Call it Q.

Step 2. Archimedes then forms the cone R whose height is p, shown in Figure 20, and whose base is a circle whose area equals the area of the curved surface of the left cone swept out by CD in Figure 18. Figure 20 shows R.

Archimedes concludes that

$$\frac{\text{area of base of } R}{\text{area of base of } Q} = \frac{\text{area of curved surface of left cone}}{\text{area of base of left cone}}$$
$$= CD/CF$$
$$= OD/p.$$

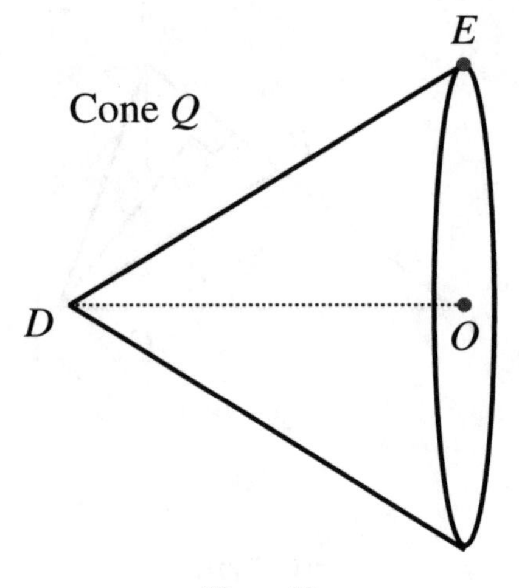

Figure 19

Thus

$$OD \cdot \text{area of base of } Q = p \cdot \text{area of base of } R.$$

Dividing both sides of this equation by 3 shows that

$$\text{volume of } Q = \frac{1}{3} \cdot p \cdot \text{area of base of } R,$$

which is the volume of R, but this fact is not needed.

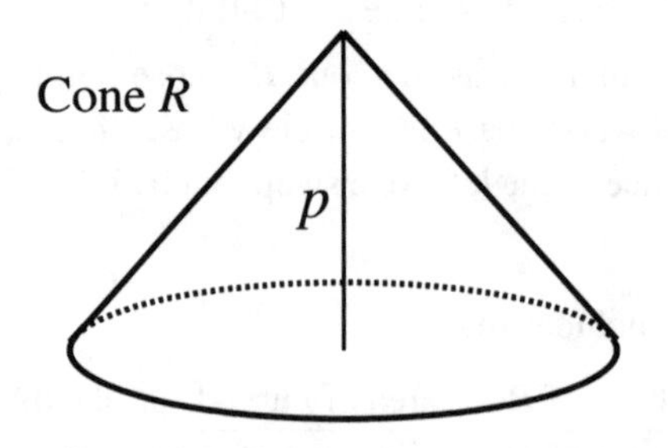

Base area equals area of curved surface of left cone in Figure 17.

Figure 20

Therefore, the volume of the double cone equals

$$\frac{1}{3} \cdot p \cdot \text{area of left surface.}$$

Finally, the volume swept out by triangle OBC of Figure 15 is the difference in the volumes of the two double cones. This difference is

$$\frac{1}{3} \cdot p \cdot \text{area swept out by triangle } OCD - \frac{1}{3} \cdot p \cdot \text{area swept out by triangle } OBD.$$

But this difference is $\left(\frac{1}{3}\right) \cdot p \cdot$ area swept out by triangle OBC.

This is precisely what the intuitive argument suggested, and completes Archimedes' derivation of the volume of a sphere as being one-third the product of its radius and its surface area.

If we look back over this chapter we see two complex and ingenious arguments. Moreover, the resulting theorems about the surface area and volume of a sphere are remarkably simple and elegant. That is the kind of discovery every mathematician and scientist hopes to make: by an involved and elegant chain of reasoning to obtain an insight that can be described in a few easily understood words. No wonder Archimedes wanted a sphere inside a cylinder placed on top of the monument over his grave. Though that monument turned to dust long ago, his discoveries will remain a part of our lives as long as civilization survives on this planet.

11

Archimedes Traps π

Today, with the use of efficient formulas for π and computers that execute a billion operations per second, it is not surprising that over fifty billion digits of π are now known. But Archimedes had no such tools. Instead he used two regular 96-sided polygons, one circumscribing the circle, one inscribing. With the aid of some elegant geometry and adroit arithmetic, he estimated the perimeters of these polygons and therefore the circumference of the circle trapped between them.

Archimedes was not thinking of the number π, but simply of the proportion "circumference : diameter." Only in the 17th century was this proportion viewed as a number and denoted by a single letter. William Jones introduced the symbol π in 1706, and by the end of the 18th century it was in common use.

Before we travel through the details of Archimedes' calculations, I will describe some of the assumptions and approximations he uses.

Background

As mentioned in the preceding chapter, Archimedes assumes that a convex curve inside the region bounded by another curve is shorter than that bounding curve. Therefore the perimeter of a polygon inscribed in a circle is less than the circumference of that circle. Similarly, the circumference of a circle is less than the perimeter of a circumscribed polygon. Thus the circumference is squeezed between the perimeters of inscribed and circumscribed polygons. By making very accurate estimates of these two perimeters in the case of regular polygons of 96 sides, Archimedes obtains close bounds on the circumference, hence on π. That, in a nutshell, is his approach.

At the beginning of his computations he needs a rational number m/n, no larger than $\sqrt{3}$ and close to it. Ideally, he would like positive integers m and n such that $(m/n)^2 = 3$ or, equivalently, $m^2 - 3n^2 = 0$, which is impossible since $\sqrt{3}$ is irrational. However, he comes close with $m/n = 265/153$, for $265^2 - 3 \cdot 153^2 = -2$. He does not reveal how he found this particular solution to the equation $x^2 - 3y^2 = -2$. A modern research paper would be just as laconic and would skip over this matter as irrelevant. (Historians have proposed several ways he may have obtained the numbers. See the references in Appendix D.) Note that

$$\left(\frac{265}{153}\right)^2 - 3 = \frac{-2}{153^2},$$

which shows that $265/153$ is only slightly less than $\sqrt{3}$.

Exercise 1.

(a) Prove that there are no integers m, n such that $m^2 - 3n^2 = -1$.
(Hint: Rewrite as $m^2 + 1 = 3n^2$ and consider remainders on division by 4.)

(b) What are the smallest positive integers m, n such that $m^2 - 3n^2 = -2$? The next smallest?

When using an inscribed 6-gon (usually called a hexagon), Archimedes needs a rational approximation of $\sqrt{3}$ larger than $\sqrt{3}$. For this, again without a word of explanation, he offers $1351/780$. Note that $1351^2 - 3 \cdot 780^2 = 1$. This implies that his estimate is just a bit larger than $\sqrt{3}$ since

$$\left(\frac{1351}{780}\right)^2 - 3 = \frac{1}{780^2}.$$

When dealing with both types of estimates Archimedes needs a relation between the dimensions of one regular polygon and the dimensions of another with twice as many sides. The key is a theorem in Book 6 of Euclid, written a generation or two before Archimedes:

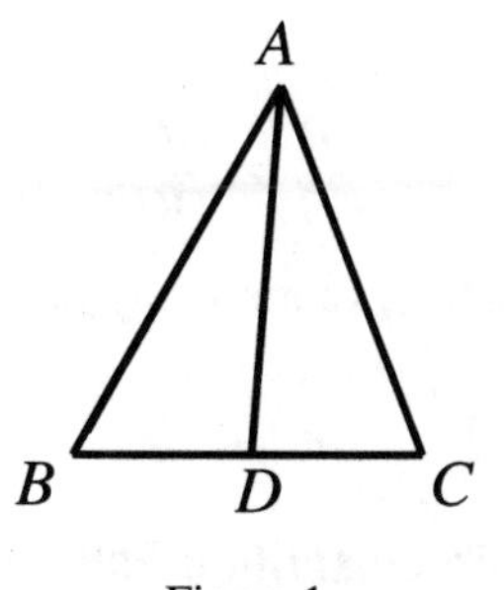

Figure 1

In triangle ABC, shown in Figure 1, line AD bisects the angle at A. Then,

$$\frac{DB}{AB} = \frac{DC}{AC}.$$

The proof is short and elegant:

Construct the line through C parallel to AD. This line meets the line AB, extended, at a point E, as shown in Figure 2.

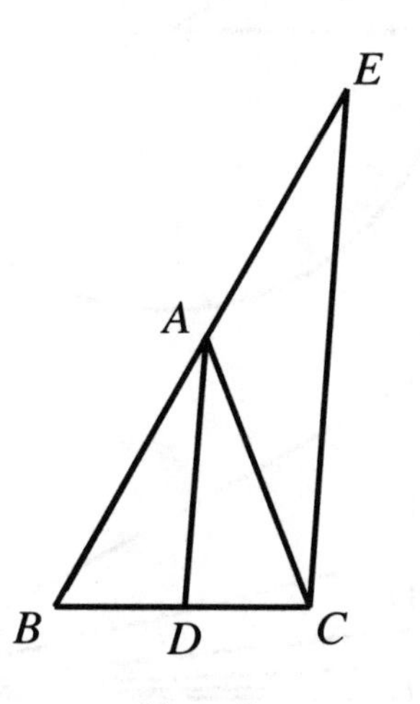

Figure 2

A quick check of angles shows that triangle CAE is isosceles. Since AD is parallel to EC,

$$\frac{AB}{DB} = \frac{AE}{DC}.$$

Hence,

$$\frac{AB}{DB} = \frac{AC}{DC}.$$

With these tools now available, we are ready to follow Archimedes' calculations.

Overestimates of π by Circumscribed Polygons

Archimedes' construction of circumscribing polygons starts with a 30° angle and then uses four successive bisections of that angle, getting the points D, E, F, and G on a tangent to the circle, as in Figure 3.

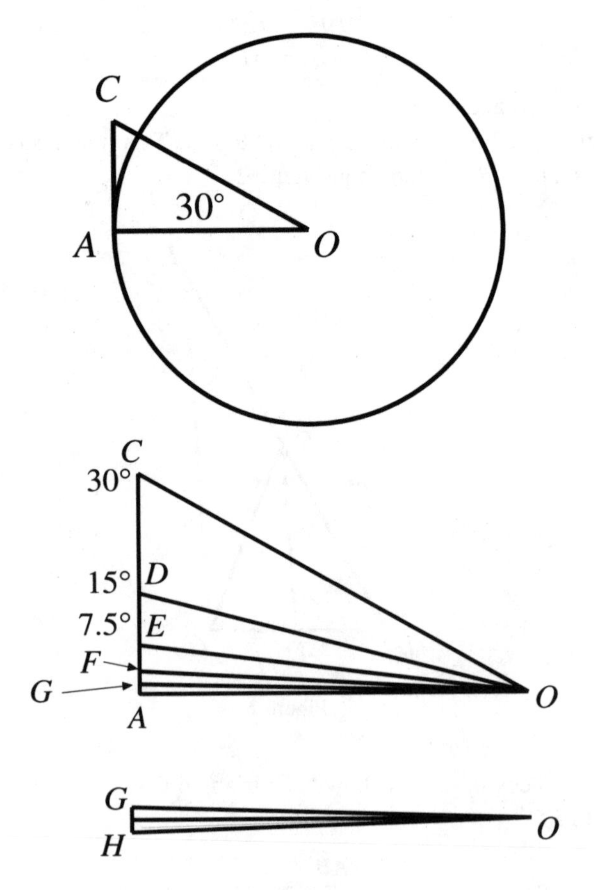

Figure 3

AC is half the side of a regular 6-gon circumscribing the circle. Similarly, AD is half the side of a regular 12-gon, and so on, until we reach AG, which is half the side of a regular 96-gon, the one that interests Archimedes. Let H be the reflection of G across the radius OA. The segment GH is one side of the regular 96-gon whose perimeter Archimedes estimates. As we carry out his computations we will, out of curiosity, see what estimate of π each of the polygons gives along the way.

The Circumscribed 6-gon

Just to get a feeling for Archimedes' approach, let us see what the circumscribed hexagon corresponding to the 30°-angle AOC tells us about π. We have

$$\frac{AO}{AC} = \sqrt{3} > \frac{265}{153}.$$

So

$$\frac{AC}{AO} < \frac{153}{265}.$$

Since the perimeter of the hexagon is $6 \cdot 2AC$ and the diameter is $2AO$,

$$\pi = \frac{\text{perimeter of circle}}{\text{diameter}} < \frac{12AC}{2AO} = 6\frac{AC}{AO} < 6 \cdot \frac{153}{265} \approx 3.464.$$

In short, π is less than 3.464.

The Circumscribed 12-gon

To get a better upper bound on π, we go to the 12-gon, each of whose edges is twice as long as AD. See Figure 4. In order to estimate AD/AO we will need not only AO/AC but also OC/AC, which is 2. For our purposes we write

$$\frac{OC}{AC} = \frac{306}{153}.$$

This is where Euclid's theorem enters the scene. In Figure 4, OD bisects angle AOC.

Therefore

$$\frac{OC}{OA} = \frac{CD}{AD}.$$

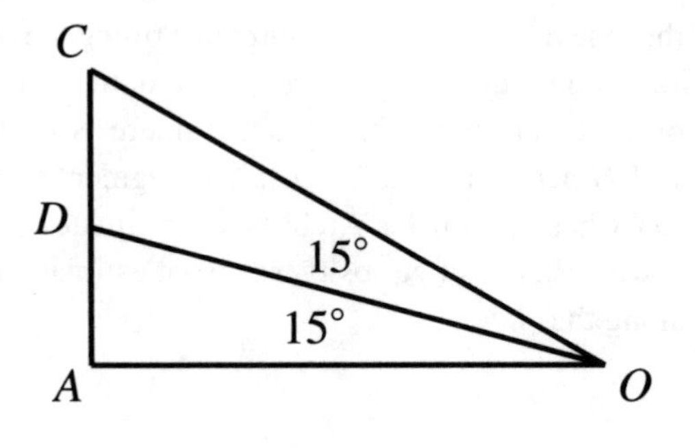

Figure 4.

Furthermore,

$$\frac{OA}{OA} = \frac{AD}{AD}.$$

Adding these two equations gives

$$\frac{OC + OA}{OA} = \frac{CD + AD}{AD};$$

hence,

$$\frac{OC + OA}{OA} = \frac{AC}{AD}.$$

Consequently,

$$\frac{OA}{AD} = \frac{OC + OA}{AC}.$$

Therefore,

$$\frac{OA}{AD} = \frac{OC + OA}{AC} = \frac{OC}{AC} + \frac{OA}{AC}.$$

But

$$\frac{OC}{AC} + \frac{OA}{AC} > \frac{306}{153} + \frac{265}{153} = \frac{571}{153},$$

so he has

$$\frac{OA}{AD} > \frac{571}{153}.$$

Exercise 2. Use the last inequality to show that π is less than $\frac{1836}{571} \approx 3.215$.

Just as we used OC/AC to help find OA/AD, we will need our estimate of OD/AD when we estimate OA/AE.

By the Pythagorean theorem, $OD^2 = AD^2 + AO^2$; hence,

$$\frac{OD}{AD} = \sqrt{1 + \left(\frac{AO}{AD}\right)^2} > \sqrt{1 + \left(\frac{571}{153}\right)^2}.$$

Now,

$$1 + \left(\frac{571}{153}\right)^2 = \frac{153^2 + 571^2}{153^2} = \frac{23409 + 326041}{153^2}.$$

So

$$\frac{OD}{AD} > \frac{\sqrt{349450}}{153},$$

and we need a good underestimate of $\sqrt{349450}$. To find it, note that $591^2 = 349281$ and $592^2 = 350464$.

It is tempting to use 591 but Archimedes resists this temptation, for 591 is not good enough for his purposes. Instead he uses $591\frac{1}{8}$, whose square is much closer to 349450:

$$\left(591\frac{1}{8}\right)^2 = 591^2 + 2 \cdot 591 \cdot \frac{1}{8} + \left(\frac{1}{8}\right)^2 = 349428 + \frac{49}{64}.$$

Incidentally, the square root of 349450 is approximately 591.14296. So $591\frac{1}{8} = 591.125$ is accurate to four significant figures.

The Circumscribed 24-gon

The shift from the 12-gon to the 24-gon is just like the shift from the 6-gon to the 12-gon. Now D and E play the roles that C and D did in the shift from the 6-gon to the 12-gon. (See Figure 4.)

The next two shifts, from the 24-gon to the 48-gon and from the 48-gon to the 96-gon, are done the same way. In the first one, to prepare for the following case, it is necessary to estimate a square root that arises from applying the Pythagorean theorem. Though I include all the details of the various steps, the reader who is interested mainly in the idea of the argument may skip to the estimate obtained with the 96-gon.

Archimedes begins his analysis of the 24-gon by estimating OA/AE and OE/AE. To do this, he begins with the equations

$$\frac{OD}{OA} = \frac{DE}{AE} \quad \text{and} \quad \frac{OA}{OA} = \frac{AE}{AE}.$$

First of all,

$$\frac{OD + OA}{OA} = \frac{DE + AE}{AE} \quad \text{or} \quad \frac{OD + OA}{OA} = \frac{AD}{AE}.$$

Thus

$$\frac{OA}{AE} = \frac{OD + OA}{AD} > \frac{591\frac{1}{8} + 571}{153} = \frac{1162\frac{1}{8}}{153}.$$

Exercise 3. Using the fact that $2AE$ is the side of a circumscribed regular 24-gon, show that $\pi < 3.16$.

As before, Archimedes needs an estimate of OE/AE to deal with the 48-gon based on the point E. Since $OE^2 = AE^2 + OA^2$,

$$\frac{OE}{AE} > \sqrt{1 + \left(\frac{1162\frac{1}{8}}{153}\right)^2} = \sqrt{\frac{23409 + 1350534\frac{33}{64}}{153}} = \frac{\sqrt{1373943\frac{33}{64}}}{153}.$$

The square root of $1373943\frac{33}{64}$ lies between 1172 and 1173, since,

$$1172^2 = 1373584 < 1373943\frac{33}{64} < 1375929 = 1173^2.$$

Archimedes uses $1172\frac{1}{8}$, not 1172, as the underestimate. As a check, $(1172\frac{1}{8})^2 = 1373877\frac{1}{64}$. As a decimal, this estimate is 1172.125, while the square root is 1172.153, to seven significant figures. Archimedes has five-place accuracy in this estimate. In any case, he has

$$\frac{OE}{AE} > \frac{1172\frac{1}{8}}{153}.$$

The Circumscribed 48-gon

In a manner analogous to the earlier cases,

$$\frac{OE + OA}{OA} = \frac{AE}{AF};$$

so

$$\frac{OA}{AF} = \frac{OE + OA}{AE} > \frac{1172\frac{1}{8} + 1162\frac{1}{8}}{153} = \frac{2334\frac{1}{4}}{153}.$$

Exercise 4. Using the inequality just obtained, show that π is less than 3.147.

Archimedes has only one more step to go, estimating OA/AG, for which he needs an estimate of OF/AF.

He has

$$\frac{OF}{AF} = \sqrt{1 + \left(\frac{OA}{AF}\right)^2} > \sqrt{1 + \left(\frac{2334\frac{1}{4}}{153}\right)^2} = \frac{\sqrt{5472132\frac{1}{6}}}{153}.$$

The square root lies between 2339 and 2340, and Archimedes uses $2339\frac{1}{4}$, whose square is $54720901\frac{1}{6}$. Thus

$$\frac{OF}{AF} > \frac{2339\frac{1}{4}}{153}.$$

The Circumscribed 96-gon

Finally he reaches the estimate of OA/OG, the key to the perimeter of the 96-gon:

$$\frac{OA}{AG} = \frac{OF + OA}{AF} > \frac{2339\frac{1}{4} + 2334\frac{1}{4}}{153} = \frac{4673\frac{1}{2}}{153}.$$

Thus

$$\frac{AG}{OA} < \frac{153}{4673\frac{1}{2}}.$$

Since

$$\pi < \frac{96 \cdot 2 \cdot AG}{2AO},$$

Archimedes has

$$\pi < \frac{96 \cdot 153}{4673\frac{1}{2}} = \frac{14688}{4673\frac{1}{2}}.$$

Rather than ask people to memorize such an ungainly fraction, Archimedes replaces it by $\frac{22}{7}$, which is only slightly larger:

$$\frac{14688}{4673\frac{1}{2}} = 3 + \frac{667\frac{1}{2}}{4673\frac{1}{2}} < 3 + \frac{667\frac{1}{2}}{4672\frac{1}{2}} = 3 + \frac{1}{7}.$$

With this he obtains the famous overestimate of π, $\frac{22}{7}$.

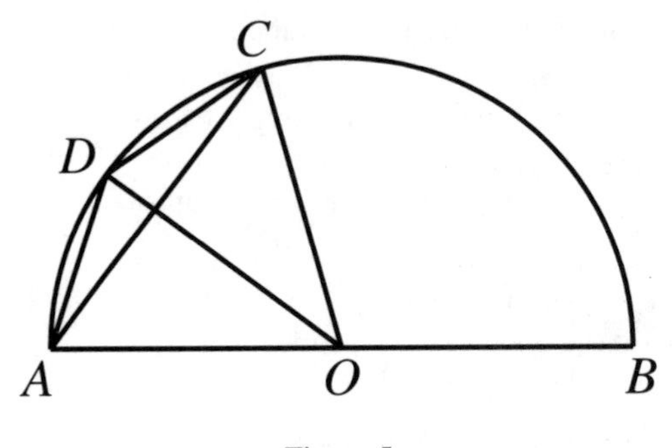

Figure 5

Underestimates of π by Inscribed Polygons

We would expect Archimedes to start with Figure 5, where D is the midpoint of the arc AC.

If chord AC is the side of an inscribed regular n-gon, then chord AD is the side of an inscribed regular $2n$-gon. However, he chooses the vertex of his angles to be B, the end of the diameter through A and O, not O, in order to exploit the similar right triangles in Figure 6.

Triangles PCB and ADB are similar (for both are right triangles and angle DBA equals angle CBD, since they are inscribed in equal arcs). Therefore

$$\frac{BD}{AD} = \frac{BC}{PC}.$$

But again by the theorem of Euclid,

$$\frac{BC}{PC} = \frac{AB}{AP}.$$

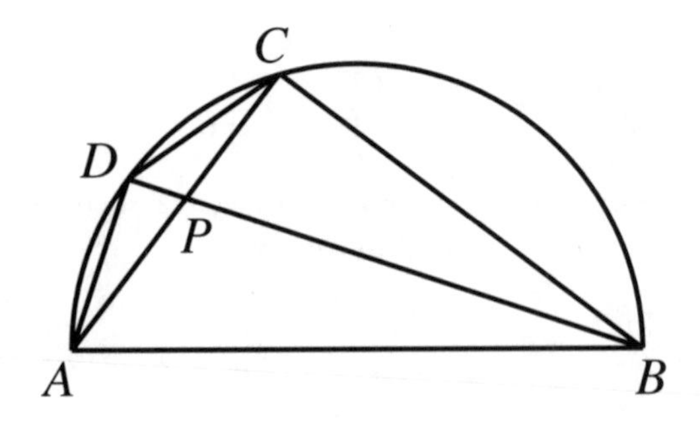

Figure 6

Thus,

$$\frac{BD}{AD} = \frac{AB}{AP}.$$

Now, it is a property of numbers that if $a/b = c/d$, then

$$\frac{a}{b} = \frac{a+c}{b+d}.$$

Exercise 5. Prove that the property just mentioned holds.

It follows that

$$\frac{BD}{AD} = \frac{AB+BC}{AP+PC} = \frac{AB+BC}{AC}.$$

This gives Archimedes his key that relates the inscribed $2n$-gon to the inscribed n-gon:

$$\frac{BD}{AD} = \frac{AB+BC}{AC}.$$

Knowing AB/AC and BC/AC for the n-gon gives BD/AD for the $2n$-gon. Then the Pythagorean theorem gives AB/AD for the $2n$-gon:

$$\frac{AB}{AD} = \frac{\sqrt{AD^2 + BD^2}}{AD} = \sqrt{1 + \left(\frac{BD}{AD}\right)^2}.$$

Repeated use of the equations for BD/AD and AB/AD, using the point E that bisects the arc AD, F that bisects the arc AE, and so on provides Archimedes with a lower estimate of π.

The Inscribed 6-gon

As with circumscribing polygons, Archimedes begins with a 6-gon, as in Figure 7.

He starts with $BC/AC = \sqrt{3}$ and $AB/AC = 2$.

Exercise 6. Show that the inscribed hexagon already tells us that $\pi > 3$.

The Inscribed 12-gon

In Figure 8, the point D bisects the arc AC, so AD is the side of an inscribed regular 12-gon.

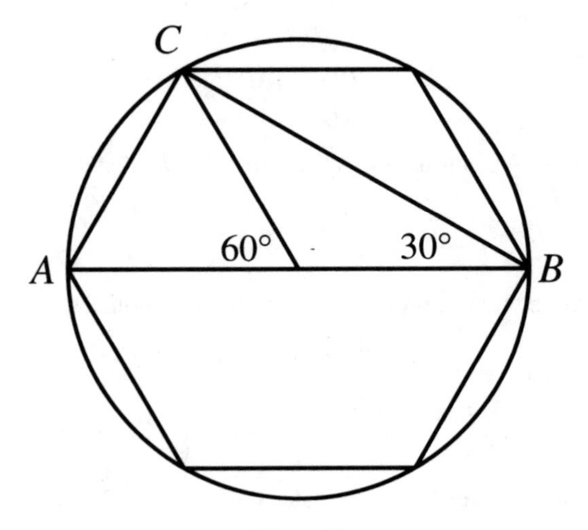

Figure 7

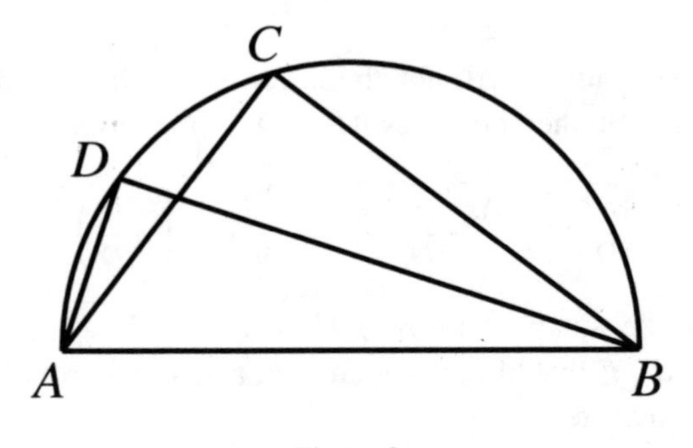

Figure 8

Since $AB/AC = 1560/780$ and $BC/AC < 1351/780$ (for $\sqrt{3} < 1351/780$, as noted on page 104), Archimedes concludes that

$$\frac{BD}{AD} < \frac{1560}{780} + \frac{1351}{780} = \frac{2911}{780}.$$

Then

$$\frac{AB}{AD} = \sqrt{1 + \left(\frac{2911}{780}\right)^2} = \frac{\sqrt{9082321}}{780} < \frac{3013\frac{3}{4}}{780}.$$

Exercise 7. Use the information just obtained to find an under-estimate of π.

The Inscribed 24-gon

The transitions between the remaining inscribed polygons are similar to the one just treated. Again, the reader interested mainly in the idea of the argument may skip the details and go to the 96-gon.

The 24-gon is determined by taking the midpoint, E, of the arc AD in Figure 8. Then AE is an edge of the inscribed regular 24-gon. The computations are like those for the 12-gon:

$$\frac{BE}{AE} = \frac{AB}{AD} + \frac{BD}{AD} < \frac{3013\frac{3}{4}}{780} + \frac{2911}{780} = \frac{5924\frac{3}{4}}{780}$$

$$= \frac{5924\frac{3}{4}}{780} \cdot \frac{\frac{4}{13}}{\frac{4}{13}} = \frac{\frac{23699}{13}}{240} = \frac{1823}{240}.$$

Then

$$\frac{AB}{AE} < \sqrt{1 + \left(\frac{1823}{240}\right)^2} = \frac{\sqrt{3380929}}{240} < \frac{1838\frac{9}{11}}{240}.$$

Exercise 8. To how many significant figures do $1838\frac{9}{11}$ and the square root of 3380929 agree?

Exercise 9. Use the last inequality to obtain an underestimate of π.

The Inscribed 48-gon

For the 48-gon, whose side is AF, the computations run the same way:

$$\frac{BF}{AF} < \frac{1823 + 1838\frac{9}{11}}{240} = \frac{3661\frac{9}{11}}{240} = \frac{3661\frac{9}{11} \cdot \frac{11}{10}}{240 \cdot \frac{11}{10}} = \frac{1007}{66}.$$

Also,

$$\frac{AB}{AF} < \sqrt{1 + \left(\frac{1007}{66}\right)^2} = \frac{\sqrt{1018405}}{66} < \frac{1009\frac{1}{6}}{66}.$$

Exercise 10. Use the last inequality to obtain an underestimate of π

The Inscribed 96-gon

At last Archimedes arrives at the 96-gon, the one he uses to get an underestimate of the circumference. Its edge is AG. The inequalities are found as before:

$$\frac{BG}{AG} < \frac{1007 + 1009\frac{1}{6}}{66} = \frac{2016\frac{1}{6}}{66},$$

and then the key inequality,

$$\frac{AB}{AG} < \sqrt{1 + \left(\frac{2016\frac{1}{6}}{66}\right)^2} = \frac{\sqrt{4069284\frac{1}{36}}}{66} < \frac{2017\frac{1}{4}}{66}.$$

Therefore the ratio between the perimeter of the inscribed regular 96-gon and the diameter AB is greater than

$$\frac{96 \cdot 66}{2017\frac{1}{4}} = \frac{6336}{2017\frac{1}{4}} = 3 + \frac{284\frac{1}{4}}{2017\frac{1}{4}} > 3\tfrac{10}{71}.$$

His underestimate $3\frac{10}{71}$ is only a little less than his overestimate $3\frac{1}{7}$, which for comparison can be expressed as $3\frac{10}{70}$.

Archimedes Traps π

Combining the information from the circumscribed and inscribed 96-gons, Archimedes obtains

$$3\tfrac{1}{7} > \frac{14688}{4673\frac{1}{2}} > \pi > \frac{6336}{2017\frac{1}{4}} > 3\tfrac{10}{71}.$$

Using only the first six digits beyond the decimal point in the decimal representation of these five numbers, we have

$$3.142857 > 3.142827 > 3.141593 > 3.140910 > 3.140845.$$

So Archimedes, by working to four-place and five-place accuracy, is able to get the first three decimal digits of π correct.

With larger denominators, he could obtain better estimates. The smallest denominator that produces an overestimate closer than $\frac{22}{7}$ to π is 113, with the estimate, $\frac{355}{113}$, which is about 3.145929. (This is also the closest estimate with a denominator at most 113.) Similarly, the smallest denominator that produces an underestimate closer than $3\frac{10}{71}$ to π is 78, with the estimate $\frac{245}{78}$, which is about 3.1410256.

A Backward Glance

Archimedes' overestimate, 3.14282, is about 0.0012 too large, and his underestimate, 3.14090, is about 0.0007 too small. That the underestimate is closer suggests that the perimeter of the inscribed 96-gon provides a better estimate of π than does the circumscribed 96-gon. Let's check this and also see how close Archimedes' estimates of the perimeters are in order to get a deeper understanding of Archimedes' approach.

The perimeter of a regular n-gon circumscribed around a circle of diameter 1 is $n \cdot \tan(\pi/n)$, and the perimeter of the inscribed regular n-gon is $n \cdot \sin(\pi/n)$. When n is 96, these formulas give perimeters of approximately 3.1427146 and 3.141032. Compare these with Archimedes' corresponding estimates, 3.14282 and 3.1409. Both are quite close to the perimeters, being off by about 0.0001 in each case. That tells us that the difference between his over- and underestimates is due mainly to the difference between the perimeters of the two polygons, not to the arithmetical errors introduced by his approximations of the various square roots.

Exercise 11. (This exercise uses calculus.)

(a) Show that the error in using the circumscribed regular n-gon to estimate π is $n \cdot \tan(\pi/n) - \pi$.

(b) Using the first two nonzero terms for the power series for $\tan x$, show that the error is roughly $\pi^3/(3n^2)$.

(c) Using l'Hopital's rule, show that

$$\frac{n \tan \frac{\pi}{n} - \pi}{\pi^3/3n^2}$$

approaches 1 as n increases. Hint: Replace $1/n$ by x.

Exercise 12. (Calculus.) As in the preceding exercise, examine the error involved in using inscribed regular n-gons to estimate π. In particular, show that when n is large, this error is about $\pi^3/(6n^2)$, which is half the estimate in the preceding exercise.

Exercise 13.

(a) In view of the preceding two exercises, what combination of the perimeters of circumscribed and inscribed n-gons would probably give the best estimate of the circumference?

(b) Use this combination with $n = 96$ to see how close it is.

What Did Archimedes Do Besides Cry Eureka?

With this chapter, in which Archimedes essentially finds the first three decimal digits for π, we come to the end of our survey of his discoveries. His investigation of the lever and center of gravity founded the science of mechanics. His research into the equilibrium of floating bodies is the first recorded work in hydrostatics and naval architecture. His determination of the areas and volumes of such familiar regions as the parabola and the sphere represents a spectacular advance in pure geometry. It is astonishing that he could do so much with the few tools available to him over two millennia ago. What miracles would he accomplish if he were born today and had at his disposal all the wondrous mathematical structures and tools developed in the last four centuries?

A

Appendix A
Affine Mappings and the Parabola

Certain mappings of the xy-plane into itself enable us to obtain easily the properties of a parabola that Archimedes uses. The proofs made with their aid are quite different from his or Euclid's. However, they have two advantages: they are short and they demonstrate the power of mappings.

The first part of this appendix develops the basic properties of these mappings; the second applies them to the parabola. We describe similar applications to the paraboloid, but omit the details.

Mappings

A *mapping* of the xy-plane into itself is a way of assigning to each point in that plane a point in the plane. Call the mapping T. If T assigns to the point (x, y) the point (x', y') write $T(x, y) = (x', y')$. We also say that T "takes the point (x, y) to the point (x', y')." The point (x, y) is called the *argument* or *pre-image* and the point (x', y') is called the *image*.

Example 1. Assign to each point the point three units to its right. What is the formula for this mapping?

Solution. In this case $x' = x + 3$ and $y' = y$. The mapping, T, in this case is given by $T(x, y) = (x + 3, y)$.

Example 2. What is the formula for the mapping T that moves each point three units to the right and two units down?

Solution. In this case $x' = x + 3$ and $y' = y - 2$. Thus $T(x, y) = (x + 3, y - 2)$.

Let h and k be constants. A mapping of the form $T(x, y) = (x + h, y + k)$ is called a "translation." It assigns to each point the point h units to the right if h is positive ($|h|$ units to the left if h is negative) and k units up if k is positive ($|k|$ units down if k is negative).

A mapping T indirectly assigns to each set S in the xy-plane an image $T(S)$ that consists of all the points $T(x, y)$ for points (x, y) in S.

Example 3. Let $T(x, y) = (x + 1, 2y + 1)$. Find the effect of T on the line $x = 3$.

Solution. Every point on the line $x = 3$ is moved to a point whose x-coordinate is $3 + 1 = 4$. Thus the image of the line $x = 3$ is on the line $x = 4$. Since $y' = 2y + 1$, if y changes by some amount, y' changes by twice that amount. Thus T magnifies segments on the line by a factor of 2.

Exercise 1. For the mapping in Example 3 compute and plot $T(x, y)$ for (x, y) equal to

(a) $(3, 1)$,

(b) $(3, 2)$,

(c) $(3, 3)$.

Example 4. Let $T(x, y) = (x', y') = (2x, 3y)$. What is the image of the circle $x^2 + y^2 = 1$?

Solution. Solve the equations $x' = 2x$ and $y' = 3y$ for x and y in terms of x' and y':

$$x = x'/2 \quad \text{and} \quad y = y'/3.$$

Combine this information with the equation that x and y satisfy to obtain an equation relating x' and y':

$$\left(\frac{x'}{2}\right)^2 + \left(\frac{y'}{3}\right)^2 = 1.$$

Among the points on this curve are $(2, 0)$, $(-2, 0)$, $(3, 0)$, and $(-3, 0)$. The curve is the ellipse shown in Figure 1.

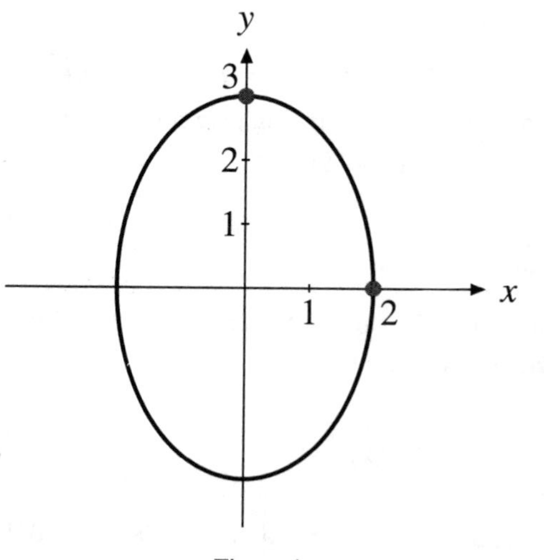

Figure 1

A mapping of the form $T(x, y) = (kx, ky)$, where k is a nonzero constant, is an example of a "similarity." If k is positive, T assigns to each point (x, y) the point k times as far from the origin $(0, 0)$ as (x, y) and on the same ray from the origin.

Example 5. What is the image of the parabola $y = x^2$ under the similarity $T(x, y) = (kx, ky)$?

Solution. Since $x' = kx$ and $y' = ky$, it follows that $x = x'/k$ and $y = y'/k$. Thus

$$\frac{y'}{k} = \left(\frac{x'}{k}\right)^2;$$

hence,

$$y' = \frac{1}{k}(x')^2.$$

Thus the image of a parabola is again a parabola. Any parabola, $y' = a(x')^2$, can be obtained this way. That means that all parabolas are similar, that is, "have the same shape." The theorems that we prove about chords, tangents, and sections of one parabola hold for all parabolas, since these concepts are preserved by similarities.

Exercise 2.

(a) Plot three points on the parabola $y = x^2$.
(b) Plot their images under the mapping given by $T(x, y) = (2x, 2y)$.
(c) Plot the image of the parabola in (a).

Affine Mappings

Translations and similarities are special cases of "affine" mappings. An *affine mapping* is described by equations of the form

$$\begin{aligned} x' &= ax + by + e \\ y' &= cx + dy + f. \end{aligned} \tag{1}$$

Here a, b, c, d, e,and f are constants and $ad - bc$ is not 0. Some of these six constants may be 0.

Exercise 3. Using algebra, show that if T is an affine mapping, then every point in the plane is the image of some point. (This is one reason the condition $ad - bc \neq 0$ is needed.)

Exercise 4. Using algebra, show that if T is an affine mapping, then the images of distinct points are distinct. (Again the condition $ad - bc \neq 0$ is needed.)

Exercise 5. Solving equations (1) for x and y in terms of x' and y' shows that

$$\begin{aligned} x &= \frac{d}{ad - bc}x' - \frac{b}{ad - bc}y' + \frac{bf - ed}{ad - bc} \\ y &= \frac{-c}{ad - bc}x' + \frac{a}{ad - bc}y' + \frac{ce - af}{ad - bc}. \end{aligned} \tag{2}$$

Supply the omitted algebra.

It is not important to memorize the formulas in Exercise 5, but it is important to note that the equations describe an affine mapping that assigns to the point (x', y') its pre-image (x, y).

Exercise 6. Check that the formulas (2) in Exercise 5 describe an affine mapping.

Formulas (2) have the form

$$x = px' + qy' + t$$
$$y = rx' + sy' + u, \qquad (3)$$

where p, q, r, s, t, and u are constants such that $ps - qr$ is not 0. This will be used in the proof of Theorem 1.

The Behavior of Lines

The most important property of any affine mapping is that it takes lines to lines, as Theorem 1 asserts.

Theorem 1. Let T be an affine mapping and L a line. Then $T(L)$ is also a line.

Proof. L has the equation $Ax + By + C = 0$, where A, B, and C are constants and not both A and B are 0. We use formula (3) to obtain an equation satisfied by points (x', y') on $T(L)$, the image of L:

$$A(px' + qy' + t) + B(rx' + sy' + u) + C = 0,$$

which can be rewritten as

$$(Ap + Br)x' + (Aq + Bs)y' + At + Bu + C = 0.$$

This last equation is the equation of a line unless the two coefficients, $Ap + Br$ and $Aq + Bs$, are both 0. That they are not both 0 is the subject of the next exercise.

Exercise 7. Show that the numbers $Ap + Br$ and $Aq + Bs$ are not both 0.

Exercise 8. Consider the affine mapping given by $T(x, y) = (2x + y, x - y)$.

(a) Draw the images of the x- and y-axes.

(b) The two axes are perpendicular. Are their images? (Suggestion: To draw the image of a line, first plot the images of two points on the line, then draw the line they determine.)

Theorem 2. An affine mapping takes parallel lines to parallel lines.

Proof. Let L_1 and L_2 be two parallel lines in the xy-plane and let L_1' and L_2' be their images. (We assume that L_1 and L_2 are not the same line.) If these two latter lines meet, there is a point P common to both. The pre-image of P would lie on both L_1 and on L_2. However, since these two lines are parallel and distinct, they share no point. This contradiction proves the theorem.

It follows from Theorem 2 that the image of a parallelogram under an affine mapping is again a parallelogram.

The Behavior of Length

Theorem 3. If two line segments are parallel and of equal lengths, so are their images under an affine mapping T.

Proof. Consider first the case where the two given segments do not lie on the same line. They form opposite sides of a parallelogram. Therefore their images under T are also opposite sides of a parallelogram. Hence they are of equal lengths and, as we already knew from Theorem 2, parallel.

Now consider the case where the two segments S_1 and S_2 lie on the same line. Introduce a third segment, S, parallel to them but not on the same line, and of the same length, as in Figure 2.

The images of the two original segments have the same length as the image of S and are parallel. This completes the proof.

Theorem 3 implies that under an affine mapping, T, "midpoints go to midpoints." In detail: Let M be the midpoint of the segment AB. Then the segments AM and MB have the same lengths. Thus their images have the same lengths. This means that $T(M)$ is the midpoint of the segment whose ends are $T(A)$ and $T(B)$.

Now let $P, Q, R, S, \ldots$ be equally spaced points on a line and T an affine mapping. Then $T(P), T(Q), T(R), T(S), \ldots$ also lie on a line. Moreover, since Q is the midpoint of PR, $T(Q)$ is the midpoint of the segment whose ends are

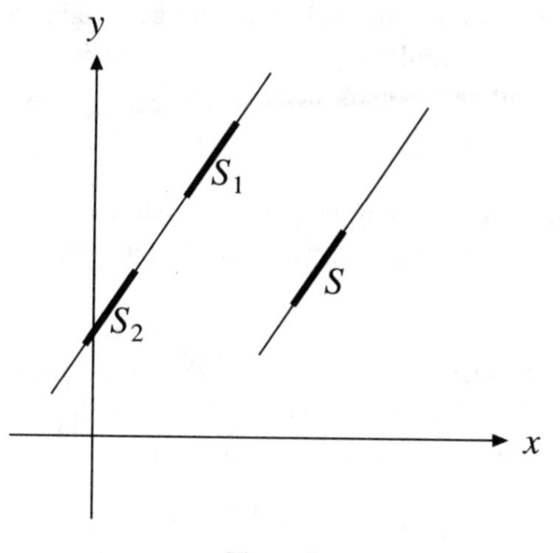

Figure 2

$T(P)$ and $T(R)$. Thus the points $T(P)$, $T(Q)$, $T(R)$ are equally spaced. It follows that an affine mapping takes equally spaced points on a line to equally spaced points (on a line), as shown in Figure 3.

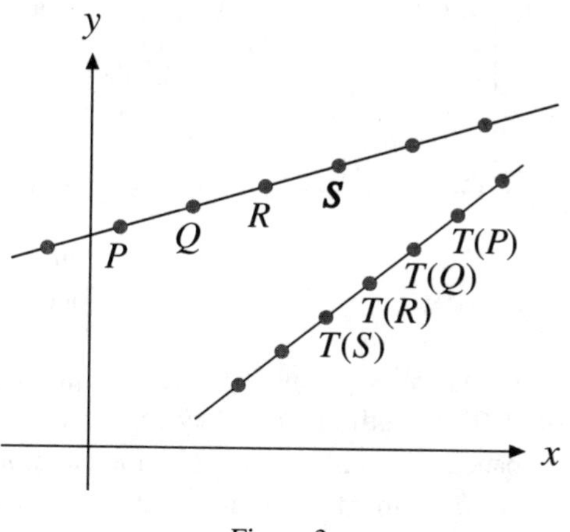

Figure 3

So far we have seen that an affine mapping preserves some geometric properties, such as "line," "parallel lines," "parallelogram," "midpoint," and "equally spaced points." However, as Examples 3 and 4 and Exercise 8 illustrate, an affine mapping need not preserve length, area, or size of angles. In particular, the image of a right angle need not be a right angle. This implies that perpendicular lines need not go to perpendicular lines.

Although an affine mapping need not preserve lengths, it does behave nicely with respect to lengths of parallel segments, as the next theorem shows.

Theorem 4. An affine mapping magnifies (or shrinks) any two parallel line segments by the same factor, which depends only on the direction of the segments.

Proof. Consider segments on a line L. Divide L into congruent segments by equally spaced points. The image of L, the line L', is divided by the images of these points into congruent segments, as in Figure 4.

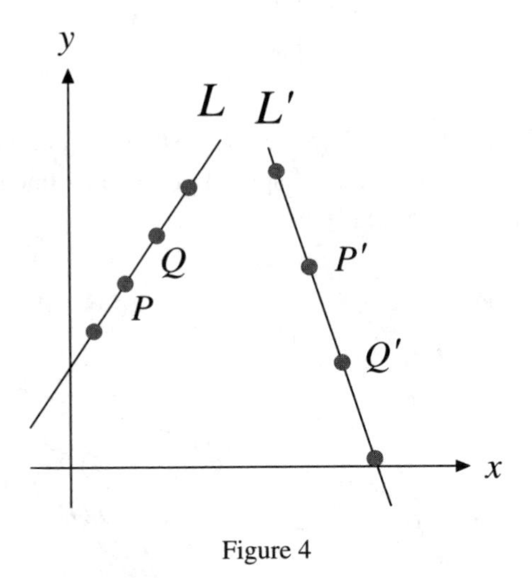

Figure 4

The image of the typical segment PQ is $P'Q'$. Let the number m be the quotient (length of $P'Q'$)/(length of PQ). Any segment on L whose ends are among the equally spaced points is magnified by the factor m.

Now subdivide line L further by introducing the midpoints of the segments into which L is divided. These, together with the original points of division, determine shorter segments, and these segments are also magnified by the same factor m. Introducing the midpoints of these shorter segments produces even shorter segments that are magnified by the factor m. By repeatedly adding midpoints we can produce arbitrarily short segments, all of which are magnified by the affine mapping by the same factor, m.

Any segment of any length can be closely approximated by a collection of very short segments. It can be shown with the aid of this information that any segment is magnified by the same factor, m.

Exercise 9. By what factor does the affine mapping $T(x, y) = (2x, y/3)$ magnify segments on

(a) the x-axis?

(b) the y-axis?

(c) the line $y = x$?

The following corollary shows that although an affine mapping doesn't preserve lengths, it does preserve the ratio of lengths of parallel segments.

Corollary. Let PQ and RS be parallel line segments. Let $P'Q'$ and $R'S'$ be their respective images under an affine mapping. Then

$$\frac{R'S'}{P'Q'} = \frac{RS}{PQ}.$$

Proof. Since an affine mapping magnifies the parallel segments PQ and RS by the same factor, say m, we have $R'S' = mRS$ and $P'Q' = mPQ$. Therefore

$$\frac{R'S'}{P'Q'} = \frac{mRS}{mPQ} = \frac{RS}{PQ}.$$

This proves the corollary.

The Behavior of Area

Because an affine mapping can magnify or shrink regions, it does not usually preserve areas. However, it does the next best thing: Each affine mapping magnifies (or shrinks) all areas by a fixed factor.

To see why, consider an affine mapping T. Cut the xy-plane into congruent squares by lines parallel to the axes, as in Figure 5. The image of each of these squares under T is a parallelogram. The parallelograms fill up the xy-plane, as in Figure 5.

Since T takes regularly spaced points on a line to regularly spaced points on a line, the parallelograms are congruent. Thus the area of each parallelogram in Figure 5 is the same multiple of the area of each square. In short, T magnifies each square by the same factor, say, s. By repeatedly dividing the squares into ever smaller squares, as we divided line segments before, we find arbitrarily

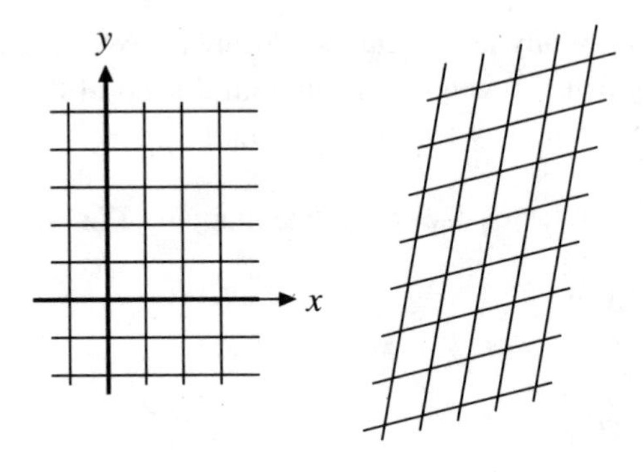

Figure 5

small squares that are magnified by the same factor, s. It can be shown, then, that T magnifies all areas—even of figures bounded by curves—by the same factor, s (since they can be approximated by collections of very small congruent squares).

Exercise 10. Find the area-magnifying factor for T given by $T(x, y) = (2x, 3y)$.

How do we compute the area-magnifying factor for a given affine mapping, $T(x, y) = (ax + by + e, cx + dy + f)$? To find out, it is enough to find how much T magnifies (or shrinks) the area of one region, say the triangle with vertices $(0, 0)$, $(1, 0)$, $(0, 1)$. That is the idea behind the proof of the following theorem.

Theorem 5. The mapping given by $T(x, y) = (ax + by + e, cx + dy + f)$ magnifies areas by $|ad - bc|$, the absolute value of $ad - bc$.

Proof. Since T takes lines to lines, the image of a triangle is a triangle. In particular, the image of the triangle with vertices $P = (0, 0)$, $Q = (1, 0)$, and $R = (0, 1)$ is the triangle with vertices $P' = (e, f)$, $Q' = (a + e, c + f)$, and $R' = (b + e, d + f)$. These are shown in Figure 6.

(In the figure R' lies above the segment $P'Q'$. Other arrangements of the three vertices, P', Q', and R', can be treated the way we deal with the arrangement in the figure.)

To find the area of the triangle $P'Q'R'$, surround it with a circumscribing rectangle with sides parallel to the axes, as in Figure 7.

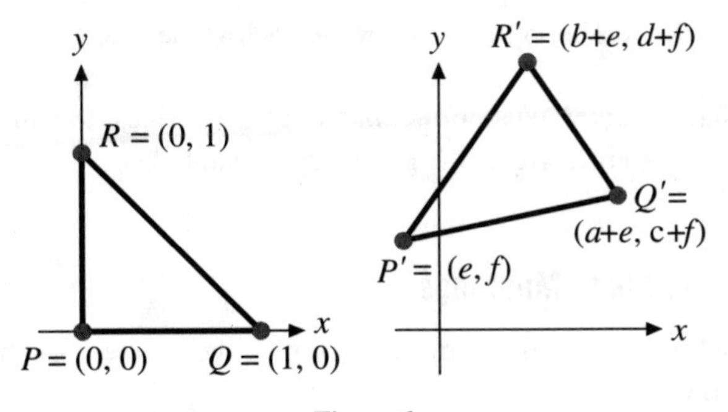

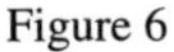
Figure 6

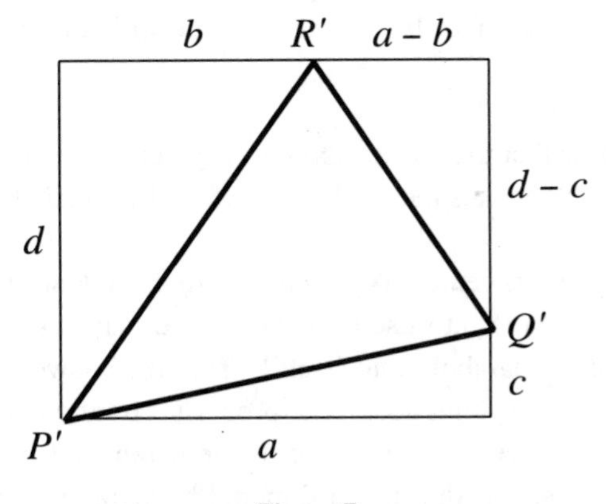

Figure 7

The dimensions of the right triangles that surround $P'Q'R'$ are shown in Figure 7. Since the area of $P'Q'R'$ is the difference between the area of the rectangle and the sum of the areas of the three triangles outside $P'Q'R'$ we have,

$$\begin{aligned} \text{area of } P'Q'R' &= ad - \left[\frac{1}{2}\cdot ac + \frac{1}{2}\cdot(a-b)(d-c) + \frac{1}{2}\cdot bd\right] \\ &= ad - \left[\frac{1}{2}\cdot ad + \frac{1}{2}\cdot bc\right] \\ &= (ad-bc)/2. \end{aligned}$$

Since the area of triangle PQR is $\frac{1}{2}$, the magnifying factor is $ad - bc$. (In some arrangements of P', Q', and R', it is the negative of $ad - bc$, when $ad - bc$ is itself negative.)

Exercise 11. Prove Theorem 5 when R' lies below the side $P'Q'$.

Now that we know what affine mappings do to lines, lengths, and areas, we can use them to investigate the geometry of parabolas.

Parabolas and Affine Mappings

Some affine mappings take some curves into themselves, as the next two exercises show.

Exercise 12. Show that the affine mapping T given by $T(x, y) = (x + 1, 2y)$ takes the curve $y = 2^x$ into itself. That is, if (x, y) is on that curve, so is its image.

Exercise 13. Show that the affine mapping T given by the formula $T(x, y) = (\frac{3}{5}x + \frac{4}{5}y, -\frac{4}{5}x + \frac{3}{5}y)$ takes the circle $x^2 + y^2 = 1$ into itself.

A rotation about its center takes a circle into itself. But a rotation certainly does not take a parabola into itself. However, we may ask, "Are there affine mappings that take a parabola into itself?" That the answer is yes gives us a powerful tool for establishing geometric properties of all parabolas.

Every parabola has the same shape as the parabola $\mathcal{P}$ whose equation is $y = x^2$, which we take as the standard model for parabolas. So we ask, "Are there affine mappings that take $\mathcal{P}$ into itself?" In other words, is there an affine mapping T such that if (x, y) is on $\mathcal{P}$, then so is $T(x, y)$? Let's see.

To find out, we simply do the necessary "brute force" calculations. Let T be given, as usual, by $T(x, y) = (ax + by + e, cx + dy + f)$. We are asking, "Are there constants a, b, c, d, e, and f such that if (x, y) is on $\mathcal{P}$, then $T(x, y)$ is also on $\mathcal{P}$?"

If the point (x, y) has the property that $y = x^2$, we want its image also to satisfy that equation. That means

$$cx + dy + f = (ax + by + e)^2.$$

Since $y = x^2$, this reduces to

$$cx + dx^2 + f = (ax + bx^2 + e)^2.$$

This equation asserts that a certain two polynomials are identical. The one on the left has degree at most 2. If b is not 0, the one on the right has degree 4.

Thus b must be 0, and we have

$$cx + dx^2 + f = (ax + e)^2.$$

Expanding the right side of the equation gives

$$cx + dx^2 + f = a^2x^2 + 2aex + e^2.$$

Comparing coefficients shows that

$$c = 2ae, d = a^2, \quad \text{and} \quad f = e^2.$$

Thus there are affine mappings of the plane that take the parabola $\mathcal{P}$ into itself. We can construct them by choosing a number a, not 0, and a number e, and letting

$$T(x, y) = (ax + e, 2aex + a^2y + e^2).$$

Exercise 14. Check that T given by $T(x, y) = (ax + e, 2aex + a^2y + e^2)$ takes an arbitrary point (p, p^2) on $\mathcal{P}$ into a point again on $\mathcal{P}$.

Exercise 15. Let $T(x, y) = (x + 3, 6x + y + 9)$.

(a) Compute and plot $T(0, 0)$, $T(1, 1)$, and $T(2, 4)$.
(b) Does T take $\mathcal{P}$ into itself?
(c) By what factor does T magnify areas?

Exercise 16. Find all affine mappings T such that $T(\mathcal{P}) = \mathcal{P}, T(1, 1) = (2, 4)$, and $T(-1, 1) = (3, 9)$.

Exercise 17. Show that the mapping given by the formula $T(x, y) = (ax + e, 2aex + a^2y + e^2)$ magnifies areas by the absolute value of a^3.

Theorem 6. Let P and Q be distinct points on the parabola $\mathcal{P}$. Let R and S also be distinct points on $\mathcal{P}$. Then there is a unique affine mapping T that takes $\mathcal{P}$ into itself and such that $T(P) = R$ and $T(Q) = S$.

Proof. Let $P = (p, p^2)$, $Q = (q, q^2)$, $R = (r, r^2)$, and $S = (s, s^2)$. We wish to prove that there are numbers a and e such that $r = ap + e$ and $s = aq + e$. These two equations can be solved for the unknowns a and e, as follows.

By subtraction, $r - s = a(p - q)$, which determines a; to be specific,

$$a = (r - s)/(p - q).$$

Substituting this value in the equation $r = ap + e$ gives an equation determining e.

Incidentally, only the parabola and the straight line have the property described in Theorem 6.

The next theorem will be helpful when using affine mappings to study parabolas.

Theorem 7. Let T be an affine mapping that takes the parabola $\mathcal{P}$ into itself. Then T takes the y-axis into a line parallel to the y-axis. Moreover, if $T(x, y) = (ax + e, 2aex + a^2y + e^2)$, then T magnifies segments of the y-axis by the factor a^2.

Proof. Since the y-axis consists of points whose x-coordinate is 0, T maps all such points to points whose x-coordinate is $a0 + e = e$. Thus the image of the y-axis is on the line $x = e$, which is parallel to the y-axis.

To determine the factor by which T magnifies (or shrinks) line segments on the y-axis, we examine the image of the segment whose ends are $(0, 0)$ and $(0, 1)$, which has length 1. Since $T(0, 0) = (e, e^2)$ and $T(0, 1) = (e, a^2 + e^2)$, its image has length $(a^2 + e^2) - e^2 = a^2$. Thus T magnifies segments of the y-axis by a factor of a^2.

As Exercise 17 shows, the mapping in Theorem 7 magnifies areas by the absolute value of a^3.

The Geometry of a Parabola

Most of the properties of a parabola that Archimedes uses involve midpoints, ratios of lengths of parallel line segments, tangents, and areas. These properties behave nicely under an affine mapping. Therefore, to establish these properties in general, it is enough to prove that they hold for particular points on the parabola $y = x^2$, which we again denote $\mathcal{P}$, and then apply a suitable affine mapping that takes $\mathcal{P}$ into itself to show that it holds for all points on $\mathcal{P}$. The proof of the following theorem illustrates the method.

Theorem 8. Let AB be a chord of a parabola and L the tangent to the parabola that is parallel to the chord. Let P be the point of tangency of L. Then the line through P parallel to the axis of the parabola bisects the chord AB and any chord parallel to AB.

Proof. It suffices to prove this for the parabola $\mathcal{P}$. Consider the simplest case, where the chord has the vertices $C = (-1, 1)$ and $D = (1, 1)$, as in Figure 8.

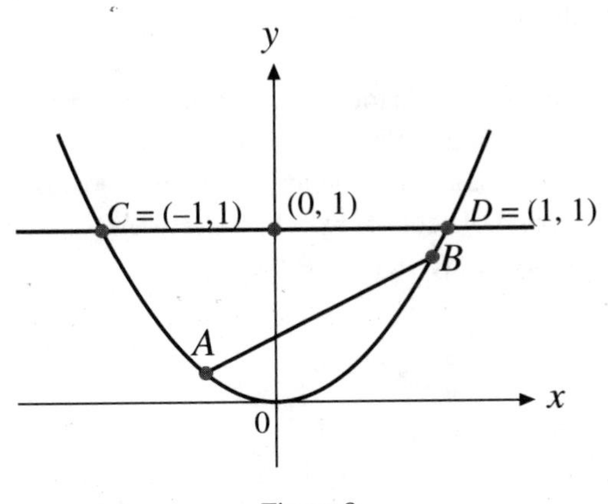

Figure 8

The tangent parallel to CD is the x-axis. The line parallel to the y-axis through the point of tangency $(0, 0)$ is the y-axis itself. Since the y-axis meets CD at its midpoint, $(0, 1)$, the theorem holds for the special case where the chord is CD.

Now let T be the affine mapping that takes $\mathcal{P}$ into itself, C to A, and D to B. T takes the chord CD to the chord AB. It takes the tangent (that is, the x-axis) to a line parallel to AB that meets the parabola at only one point. Thus the image of the x-axis is the tangent to the parabola that is parallel to AB.

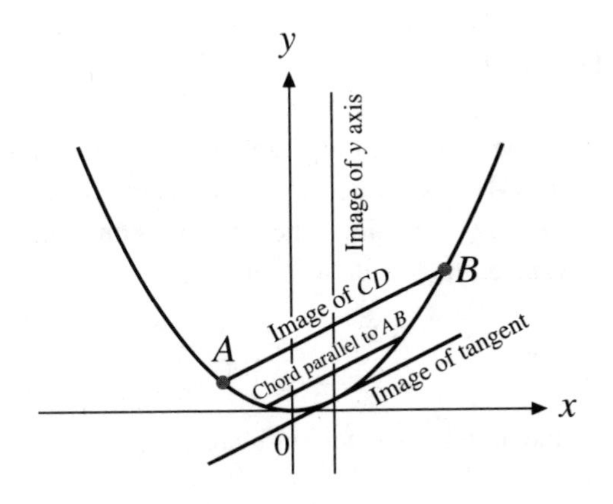

Figure 9

Moreover, it takes the y-axis to a line parallel to the y-axis. Since an affine mapping takes midpoints to midpoints, the image of the y-axis bisects all the chords parallel to AB. This is shown in Figure 9.

This proves the theorem.

Let AB be a chord of a parabola and L the line parallel to AB that is tangent to the parabola, at a point P. P is the vertex of the segment cut off by AB. The line through P parallel to the axis of the parabola is called the axis of the segment of the parabola cut off by AB. It meets AB at a point, V, which is the midpoint of AB, as shown in Figure 10.

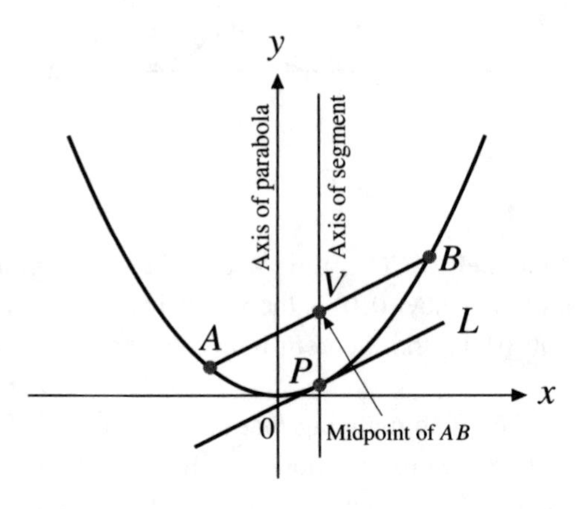

Figure 10

The next theorem can also be proved by reducing it to the case when the chord joins $(-1, 1)$ and $(1, 1)$.

Theorem 9. Let AB be a chord of a parabola and P the point of tangency of the tangent parallel to AB. Let V and V' be points on the axis of the segment cut off by AB. Let Q and Q' be points where the lines through V and V' parallel to AB, respectively, meet the parabola. Then

$$\frac{PV}{PV'} = \frac{QV^2}{(Q'V')^2}.$$

The objects in the theorem are shown in Figure 11.

Exercise 18. Prove Theorem 9.

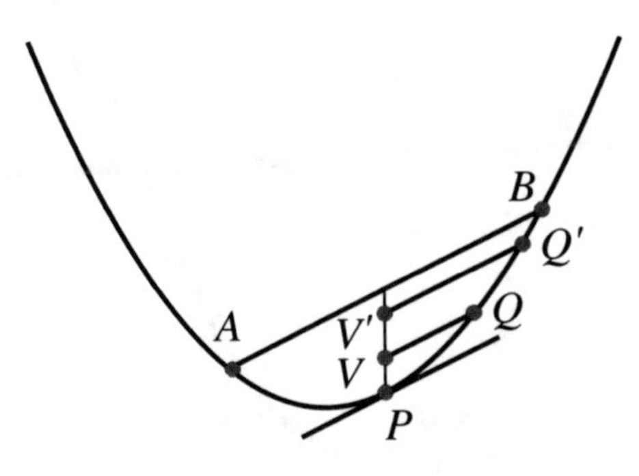

Figure 11

Exercise 19. In Figure 12, AB is a chord of the parabola and PV is the axis of the segment it determines. Prove that the area of the segment is proportional to $PV^{3/2}$. Suggestion: First look back at Theorem 7 and Exercise 17.

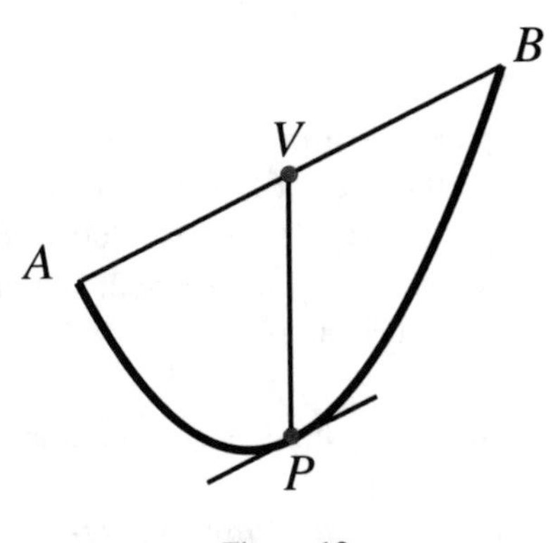

Figure 12

Exercise 20. In Figure 13, AB is a chord of the parabola, V its midpoint, P the point of tangency of the tangent parallel to AB, and Q the intersection of the tangents at A and B. Prove that V, P, and Q lie on a line.

In the next exercise the tangent to the parabola $y = x^2$ at $(1, 1)$ is needed. Though easily found with calculus, it can be found without calculus as follows. A typical line through $(1, 1)$ has slope m and equation $y - 1 = m(x - 1)$. We want the value of m such that this line meets the parabola only at the point $(1, 1)$.

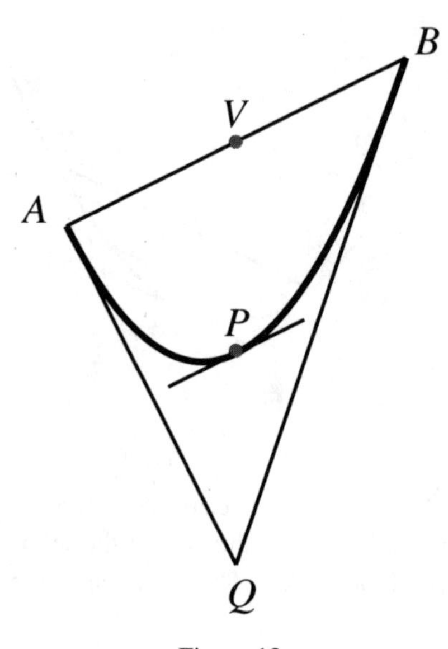

Figure 13

The line meets the parabola at (x, x^2) if $x^2 - 1 = m(x - 1)$. Since we are interested in solutions x that are not 1, we may divide both sides by $x - 1$, obtaining the equation $x + 1 = m$. If the only solution to this is $x = 1$, then m must be 2. Thus the tangent at $(1, 1)$ has slope 2, and therefore we have the equation $y - 1 = 2(x - 1)$, which simplifies to $y = 2x - 1$.

Exercise 21. In Figure 14, qQ is a chord of a parabola and QE is tangent to the parabola at Q. An arbitrary line parallel to the axis of the segment meets qQ at O, the parabola at P, and QE at E^*. Prove that $OP/OE^* = qO/qQ$. (First settle the case where $q = (-1, 1)$ and $Q = (1, 1)$ on the parabola $y = x^2$.)

Exercise 22. (This does not use affine mappings since generally a right angle is not preserved by such a mapping.) The focus of the parabola $y = x^2$ is $(0, \frac{1}{4})$. Let $A = (a, a^2)$ be any point on the parabola other than $(0, 0)$. Let L be the tangent to the parabola at A; let N be the line perpendicular to L at A. N meets the axis of the parabola at B, and the line through A perpendicular to the axis meets the axis at C. Prove that the length of BC is $\frac{1}{2}$, twice the distance from the focus to the vertex. Because all parabolas are similar, the analogous distance BC is always twice the distance from the focus to the vertex for any parabola. See Figure 15. Hint:

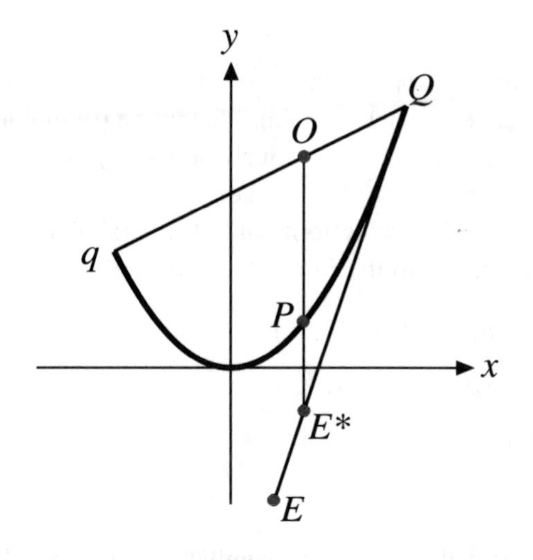

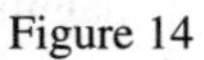
Figure 14

(a) First find the slope and equation of the tangent line at A.
(b) The slope of N is the negative of the reciprocal of the slope of the tangent. Find the equation of N.
(c) Find B and C and show that the length of BC is $\frac{1}{2}$.

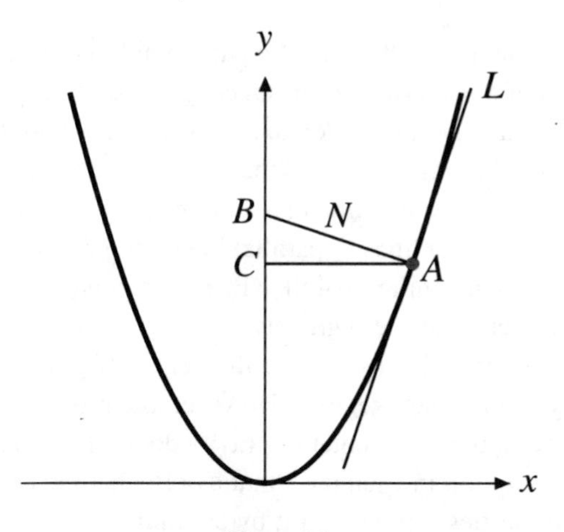

Figure 15

Affine mappings have many other applications. For instance, my article *Exactly How Did Newton Deal with His Planets?*, cited in the references, used them to obtain properties of an ellipse from properties of a circle.

The Paraboloid

Just as there are affine mappings in the xy-plane, there are affine mappings in xyz-space. There are even such mappings that take a paraboloid into itself.

If you spin the parabola $z = x^2$ around the z-axis you obtain the paraboloid whose equation is $z = x^2 + y^2$. For any constants, t, u, j, and k, where t is not 0, any mapping of the following form takes the paraboloid into itself. (It was found by brute force, as was the case with the parabola.)

$$x' = t(\cos u)x - t(\sin u)y + j$$

$$y' = t(\sin u)x + t(\cos u)y + k$$

$$z' = (2tj(\cos u) + 2tk(\sin u))x + (-2tj(\sin u) + 2tk(\cos u))y + t^2z + j^2 + k^2.$$

It can be shown that this mapping magnifies volumes by the factor t^4 and segments parallel to the z-axis by the factor t^2. This implies that the volume of a section of a paraboloid is proportional to the square of its axis, a fact that plays a key role in Chapter 8.

Exercise 23. Check that the mapping described by the equations given above takes the given paraboloid into itself.

These affine mappings that take the paraboloid into itself behave like the similar mappings of a parabola. For instance, given distinct points P and Q on the paraboloid and distinct points R and S on the paraboloid, there is an affine mapping that takes P to R, Q to S, and the paraboloid to itself.

It can be shown that any affine mapping in space magnifies volumes by a constant and takes parallel lines to parallel lines and parallel planes to parallel planes. Moreover, it magnifies parallel line segments by a constant, which depends on the direction of the segments.

An affine mapping also assigns to the center of gravity of a region the center of gravity of the image region. (However, the proof of this requires calculus, since the definition of a center of gravity does.) This suggests modifying Archimedes' assumption in Chapter 3 that asserts that the center of gravity is preserved by similarities. Strengthen it by assuming that the center of gravity is preserved by all affine mappings. At the same time weaken the assumption about the center of gravity of the union of two regions, assuming only that it lies on the line through the two given centers of gravity. Do the modified assumptions completely describe the centers of gravity of polygons?

Exercise 24. Assume that any triangle is the image of an equilateral triangle by some affine mapping. Combine this with the fact that affine mappings preserve centers of gravity, to find the center of gravity of any triangle.

Archimedes' work on the equilibrium of floating bodies was simplified by the fact that the submerged part is the affine image of the entire body. Had he allowed the base to cut across the surface of the water, this would not have been the case.

B

Appendix B
The Floating Paraboloid: Special Case

This appendix treats the case of a paraboloid floating in such a way that its exposed base just touches the water's surface. The following lemma, which Archimedes assumes is well known, is the key. The notation is that used in Chapter 8.

Lemma 1. Consider a section of a parabola with vertex A and base BC perpendicular to its axis, which meets BC at the point N. Let CD be another chord of the parabola and P the point on the parabola where the tangent is parallel to CD. Let the line through P parallel to the axis meet CD at V. Let K be a point on PV and O the intersection of AN with the line through K perpendicular to AN.

Then

$$\frac{PV}{PK} \geq \frac{ON}{OA}.$$

Proof. Figure 1 displays the data for the parabola $y = x^2$, where $K = (a, b)$ and $C = (c, c^2)$.

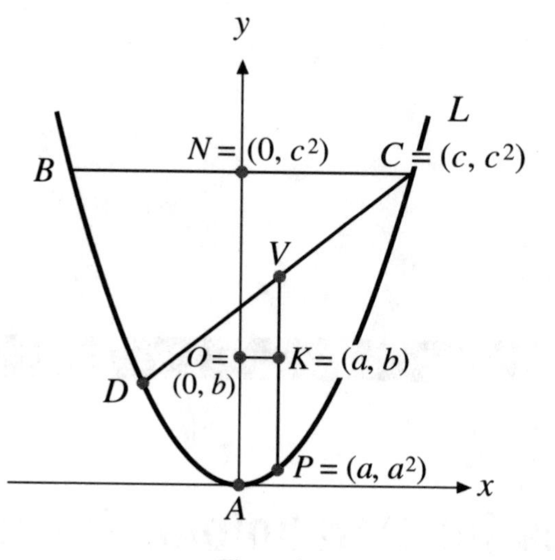

Figure 1

We have $P = (a, a^2)$ and $O = (0, b)$. In order to determine the y-coordinate of V, we first find the slope of the tangent at P. By the method used in Appendix A, this slope can be shown to be $2a$. Thus CD is part of the line whose equation is

$$\frac{y - c^2}{x - c} = 2a.$$

When $x = a$, this gives $y = c^2 + 2a^2 - 2ac$, hence $V = (a, c^2 + 2a^2 - 2ac)$.

The alleged inequality, $PV/PK = ON/OA$, translated into coordinates, becomes

$$\frac{c^2 + a^2 - 2ac}{b - a^2} \geq \frac{c^2 - b}{b}.$$

Clearing denominators and canceling changes this to

$$b^2 + a^2c^2 - 2abc \geq 0,$$

which can be written as

$$(b - ac)^2 \geq 0.$$

Since the square of a real number is greater than or equal to 0, and all the steps are reversible, the lemma is proved.

Exercise 1. Show that the tangent at $P = (a, a^2)$ has slope $2a$.

Now to the paraboloid, as shown in Figure 2. The surface of the water touches the edge of the base. We assume $AN > 3f$, since the case $AN < 3f$ has been treated in Chapter 8. The question is, "How large can AN be and still have K below F, so that there is a force tending to push the paraboloid back up?"

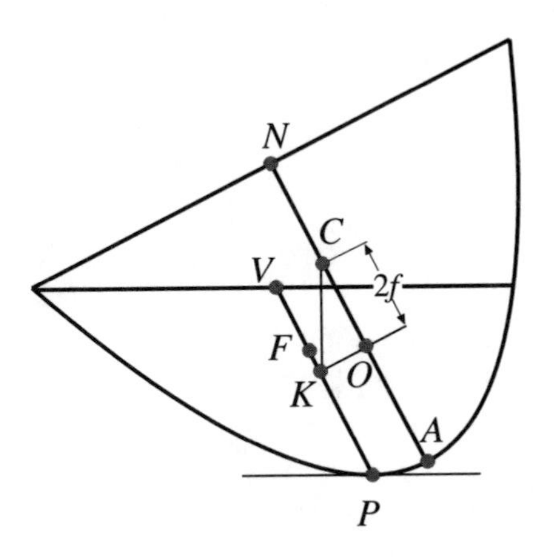

Figure 2

We wish to have PV/PK greater than PV/PF, which is $\frac{3}{2}$. In view of the lemma, it is enough to have ON/OA greater than $\frac{3}{2}$.

Since $ON = AN/3 + 2f$ and $OA = 2AN/3 - 2f$, this inequality becomes

$$\frac{\frac{AN}{3} + 2f}{\frac{2AN}{3} - 2f} > \frac{3}{2},$$

which a little algebra reduces to

$$AN < \frac{15}{2} f.$$

(This is where the number $\frac{15}{2}$ makes its appearance.) The steps are reversible. Thus, if $AN < \frac{15}{2} f$, the paraboloid in that special position tends to turn toward the vertical.

C

Appendix C
Notation

When Archimedes in Chapter 11 estimates the circumference of a circle he deals with numbers as large as 9,082,321 and fractions such as $\frac{1}{8}$ and $\frac{3}{4}$. We may well wonder how he, without the benefit of our decimal system, wrote such numbers and carried out his calculations. This appendix describes the Greek system in common use during his lifetime, a system that uses the number 10 as a base but lacks our delightful zero.

The following table shows the Greek notation for the numbers one through nine, 10 through 90 by tens, and 100 through 900 by hundreds. Each of these 27 numbers is denoted by a letter of the Greek alphabet.

1	α	(alpha)	10	ι	(iota)	100	ρ	(rho)
2	β	(beta)	20	κ	(kappa)	200	σ	(sigma)
3	γ	(gamma)	30	λ	(lambda)	300	τ	(tau)
4	δ	(delta)	40	μ	(mu)	400	υ	(upsilon)
5	ε	(epsilon)	50	ν	(nu)	500	ϕ	(phi)
6	ς	(final sigma)	60	ξ	(xi)	600	χ	(chi)
7	ζ	(zeta)	70	o	(omicron)	700	ψ	(psi)
8	η	(eta)	80	π	(pi)	800	ω	(omega)
9	θ	(theta)	90	ϙ	(koppa: archaic)	900	ϡ	(sampi: archaic)

Observe that the Greek notations for, say, 4, 40, and 400 are δ, μ, and υ, which do not display their close relation, while our notation does by use of the symbol 0. However, this difference disappears on an abacus. For instance, both the Greeks and we would represent 400 by placing four beads or pebbles in the "hundreds" column, and no beads in the tens or units columns. So their calculations with an abacus would be the same as ours.

Archimedes would write our 153 as $\rho\nu\gamma$ and 103 as $\rho\gamma$. Numbers up to 999 are easily represented in a similar way, using at most three symbols from the table.

The numbers from 1000 through 9000, by thousands, are represented by the symbols for 1 through 9 with a small vertical mark below and in front of the symbol. For instance, $_{,}\beta$ denotes 2000. The number 3013, which appears in the Chapter 11, is written $_{,}\gamma\iota\gamma$. The notations mentioned so far take us up through 9999.

Ten thousand is denoted M, for the Greek word for myriads. The number of myriads is indicated by placing appropriate symbols above the M. For instance, 20,000 is $\overset{\beta}{M}$ and 320,000 is $\overset{\lambda\beta}{M}$. Thus

$$\begin{array}{cccccc} 9{,}082{,}321 = & \overset{\text{ϡ}\eta}{M} & _{,}\beta & \tau & \kappa & \alpha \\ & 9080000 & 2000 & 300 & 20 & 1. \end{array}$$

A fraction whose numerator is one, a "unit fraction," is expressed by placing an apostrophe above and to the right of the denominator. For instance,

$$\frac{1}{3} = \gamma'$$

In general, the numerator and denominator are written on a line, separated, with an apostrophe on the denominator. For instance

$$\frac{9}{11} = \theta \quad \iota\alpha'$$

With overbars added for clarity, they write $1839\frac{9}{11}$, another number occurring in Chapter 11, as

$$\overline{.\alpha\omega\lambda\theta} \quad \overline{\theta} \quad \iota\alpha'.$$

Calculations in the Greek notation are similar to our own. In fact, on an abacus they would be the same. For instance, let's compute 24×35 in slow motion in both systems, ours and the Greek:

35			$\lambda\varepsilon$
24			$\kappa\delta$
600	100		$\chi\rho$
	120	20	$\rho\kappa\kappa$
$600 + 220 + 20 = 840.$			$\omega\kappa\kappa = \omega\mu.$

The Greek reckoner would have to memorize that $\kappa \times \lambda = \chi$ (which is 20×30), while we are obliged to learn just 2×3 and how to keep track of 0's.

Exercise 1. Compute $87 + 95$

(a) in our system, displaying the sums $80 + 90$ and $7 + 5$;

(b) in the Greek system.

Exercise 2. In Chapter 11, Archimedes meets the square of 591. Displaying all the products, compute it

(a) in our system;

(b) in the Greek system.

If we practiced arithmetic using the Greek notation for as long as we have worked with our own, perhaps the contrast between them would diminish, if not vanish.

More information about ancient arithmetic may be found in the sources cited in the references in Appendix D.

D

Appendix D References

General:

1. E. J. Dijksterhuis, *Archimedes* (with a new bibliographic essay by Wilbur R. Knorr), Princeton University Press, Princeton, 1987.
2. T. L. Heath, *The Works of Archimedes*, Dover, New York, 1953. (This is also part of Volume 11 of Britannica Great Books. The other mathematicians in this volume are Euclid, Apollonius, and Nichomachos.)
3. Paul ver Eecke, *Les Oeuvres Complétes D'Archimède*, Blanchard, Paris, 1960.
4. B. L. Van der Waerden, *Science Awakening*, P. Noordhoff Ltd., Holland, 1954, pp. 208–228.
5. C. H. Edwards, *The Historical Development of the Calculus*, Springer, New York, 1982, pp. 29–76.

Chapter 1
The Life of Archimedes?

1. Mason L. Weems, *The Life of Washington*, Marcus Cunliffe, ed., Belknap Press of Harvard University Press, Cambridge, 1962, pp. xiii–xxvi.

2. Cicero, *Tusculan Disputations*, translated by J. E. King, Loeb Classical Library, Volume 23, Harvard University Press, Cambridge, 1950, pp. 491–493.
3. Livy, *Book 24*, translated by F. G. Moore, Loeb Classical Library, Volume 6, Harvard University Press, Cambridge, 1958, pp. 282–286.
4. *Histories of Polybius*, Volume 1, Indiana University Press, Bloomington, 1962, pp. 530–533.
5. Plutarch, *Plutarch's Lives*, Loeb Classical Library, Volume 5, Life of Marcellus, Harvard University Press, Cambridge, 1968, pp. 437 and 481.
6. D. L. Simms, Archimedes and the Burning Mirrors of Syracuce, *Technology and Culture*, Volume 18, 1977, pp. 1–24.
7. P. G. Walsh, *Livy, His Historical Aims and Methods*, Cambridge University Press, Cambridge, 1961 (Chapter 2, The Tradition of Ancient Historiography).

Chapter 3
The Center of Gravity

Archimedes' definition of convexity is phrased a little differently from the one in this chapter. See *Heath: On the Sphere and Cylinder*, Book 1, definitions 2, 3, and 4 and *On the Equilibrium of Planes*, Book 1, postulate 4 or Dijksterhuis: pp. 142–145 and 287.

1. Olaf Schmidt, A System of Axioms for the Archimedean Theory of Equilibrium and Centre of Gravity, *Centaurus*, Volume 19, 1975, pp. 1–35.

Chapter 4
Big Literary Find in Constantinople

1. J. L. Heiberg, Eine Neue Archimedeshandschrift, *Hermes*, Volume 42, 1907, pp. 235–303 (Heiberg's detailed report on the lost manuscript).
2. *The Archimedes Palimpsest*, Christie's, 502 Park Avenue, New York, 10022 (a detailed analysis of the palimpsest, including many reproductions in color, some enhanced digitally or by ultraviolet light).

Chapter 6
Two Sums

1. C. H. Edwards, Jr., *The Historical Development of the Calculus*, Springer, New York, 1982 (pp. 83–85 describe how Alhazen found and applied the formulas for the sum of cubes and sum of fourth powers).

Chapter 7
The Parabola

1. Wilbur Knorr, Archimedes' Lost Treatise on the Centers of Gravity of Solids, *Mathematical Intelligencer*, Volume 1, 1978, pp. 102–109.

Chapter 8
Floating Bodies

1. G. J. Toomer, *Diocles on Burning Mirrors*, Springer, New York, 1976 (pp. 1–17 discuss the early history of conics, in particular, the focus-directrix definition of a parabola and the fact that the length OC in Figure 5 is independent of P).

Chapter 11
Archimedes traps π

1. Lennert Berggren, Jonathan Borwein, and Peter Borwein, *Pi: A Source Book*, Springer, New York, 1996.
2. David Blatner, *The Joy of π*, Walker, New York, 1997 (includes other estimates of π using polygons with 192 and 3072 sides).
3. Florian Cajori, *A History of Mathematical Notations*, Volume 2, Open Court, Chicago, 1952, pp. 8–10 (notations related to circumference of circle).
4. Dijksterhuis, pp. 229–238 (discussion of methods that Archimedes may have used to estimate square roots).
5. P. M. Gruber and J. M. Wills, *Convexity and its Applications*, Birkhäuser, Boston, 1983, p. 139 (mentions that the error in using a circumscribed polygon to estimate the perimeter of a smooth curve is approximately twice the error in using an inscribed polygon).

For information on the decimal representation of π, consult

http://www.cecm.sfu.ca/personal/jborwein/Kanada_50b.html

(In the first 50 billion decimal places, the ten digits are about equally represented. For instance, 4 occurs 5,000,023,598 times and 5 occurs 4,999,991,499 times. However, no one can prove that this trend will continue as more digits are computed.)

Appendix C
Notation

1. Dijksterhuis, pp. 55–56
2. van der Waerden, pp. 45–47
3. Thomas Heath, *A History of Greek Mathematics*, Volume 1, Oxford, 1921, pp. 29–53.
4. James Gow, *A Short History of Greek Mathematics*, Chelsea, New York, 1968, pp. 22–61.

Index